Digital Signal Processing
Lecture Notes 2021

Thomas Paul Weldon

ISBN-13: 9798534003574

CONTENTS

PREFACE

The updated collection of lecture notes in this book are based on over 20 years of teaching graduate and undergraduate courses in digital signal processing at the University of North Carolina at Charlotte. As lecture notes, this book is not intended to be a substitute for the many excellent textbooks in this field. Instead, this book is intended as a supplement to other course materials and as a workbook for students taking notes during corresponding lectures. In addition, practicing engineers may find this book useful for quick review of the topic.

Thomas Paul Weldon
Charlotte, NC
July 8, 2021

1 INTRODUCTION AND OVERVIEW

The lecture notes in this chapter provide a brief introduction to digital signal processing and an overview of the topics to be covered in the following chapters.

DSP

- Why DSP (digital signal processing)?
  - Digital circuit advantages speed/cost/density, no re-design
  - Analog circuit variation/tolerances
  - Implement adaptive systems
  - Security/encryption
  - Advances in algorithms (FFT/filters)
  - New DSP hardware capabilities in processing power

- Why not?
  - Speed limits on ADC/DAC/data-bus/etc.
  - Where do you store 1 GByte/Second?
  - Power consumption at high clock rates

- Applications:
  - Compression (save memory/bandwidth)
  - Software-defined radio (GNU radio, USRP, RFSoC)
  - Digital Non-Foster Circuits (negative capacitor, etc.)
  - HDTV
  - Cellular/ Communication/WiFi
  - Video/Image processing, JEPEG, MPEG
  - CD players
  - Security/encryption/banking
  - Digital control systems (ECGR4112)
  - Digital non-Foster

Quick overview of topics to be covered

Outline/Topics

- Discrete time signals/systems
  - Linearity, stability, causality,
  - Convolution
  - DTFT, frequency domain
  - Block diagrams

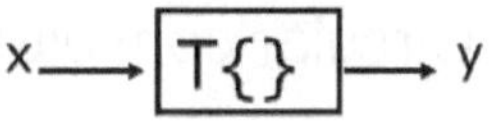

x → T{} → y

$$X(\omega) = \sum_{n=-\infty}^{\infty} x[n]e^{-j\omega n}$$

$$x[n] = \frac{1}{2\pi}\int_{-\pi}^{\pi} X(\omega)e^{j\omega n}\,d\omega$$

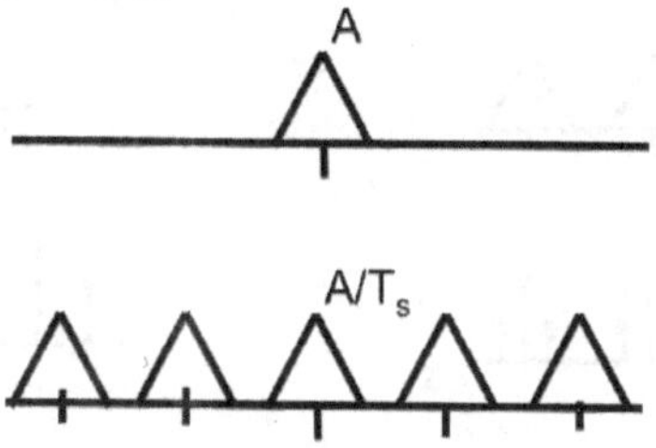

A

A/T_s

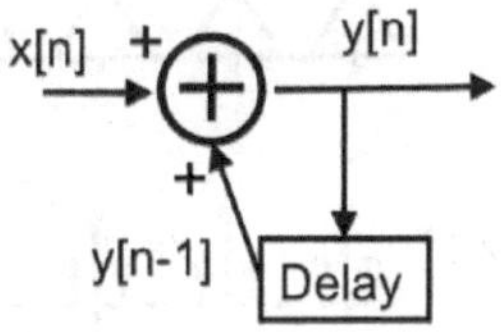

x[n] + y[n]

+

y[n-1] Delay

Outline/Topics

- Sampling of continuous signals:
  - Aliasing
  - Reconstruction
  - Quantization error

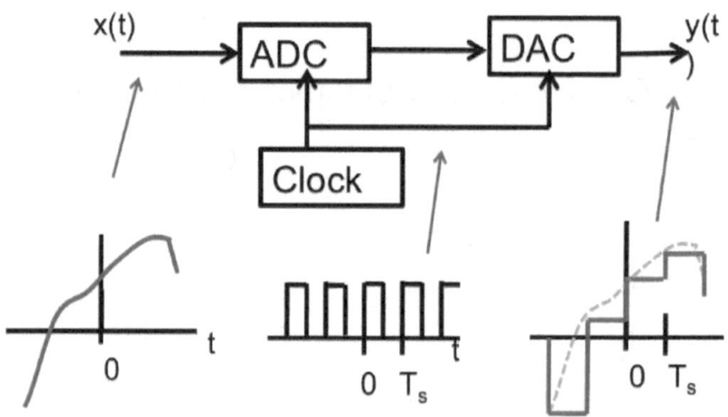

Outline/Topics

- DFT & FFT
  - DFT properties
  - FFT is a fast DFT
  - Matrix form
  - Circular convolution

DFT :

$$X[k] = \sum_{n=0}^{N-1} x[n] e^{-j\frac{2\pi}{N}nk} = X(\omega)\Big|_{\omega=2\pi k/N}$$

$$x[n] = \frac{1}{N}\sum_{k=0}^{N-1} X[k] e^{j\frac{2\pi}{N}nk}$$

$$\overline{X} = \overline{\overline{W}}\,\overline{x}$$

$$\overline{x} = \overline{\overline{W}}^{-1}\,\overline{X}$$

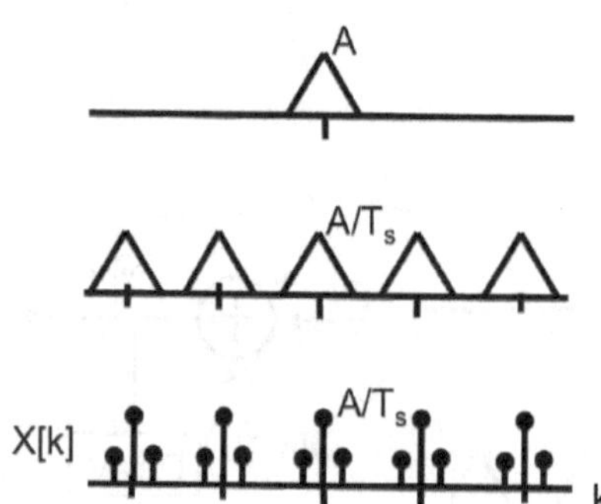

Outline/Topics

- Z-transform:
  - ROC, poles,
  - Properties,
  - Relation to DFT, DTFT
  - Freq. response,
  - Convolution,
  - BIBO stability,
  - Difference eqs.
  - Poles, zeroes

x[n]
y[n]
a
z^{-1}

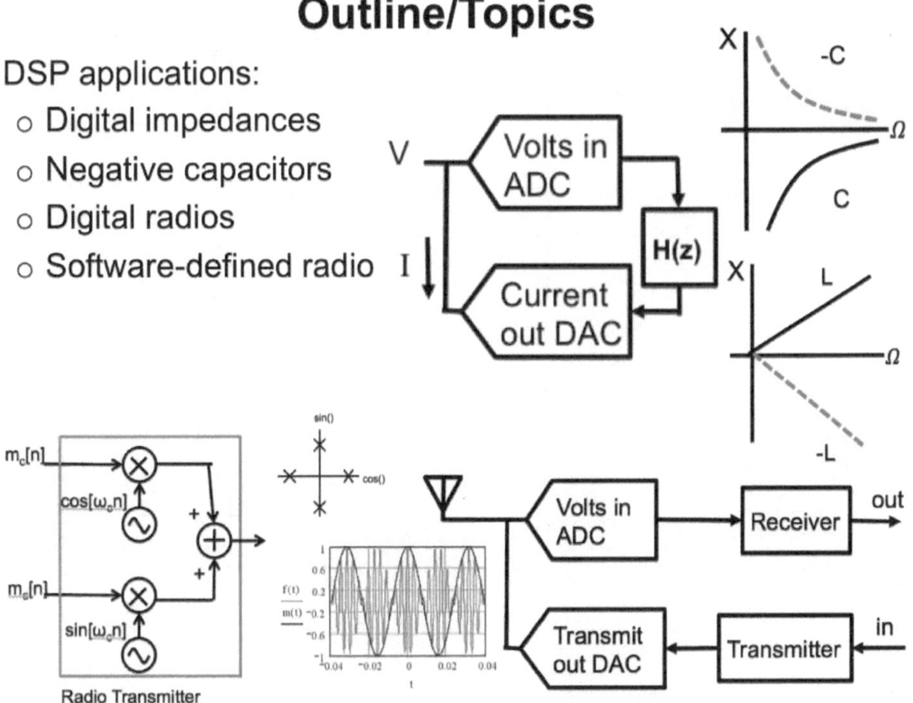

Outline/Topics

- Digital filter design:
  - IIR filters
    - Impulse invariance,
    - Bilinear transform,
  - FIR filters
    - Windowing
- Filters are not "just filters"
  - Filter = any LTI

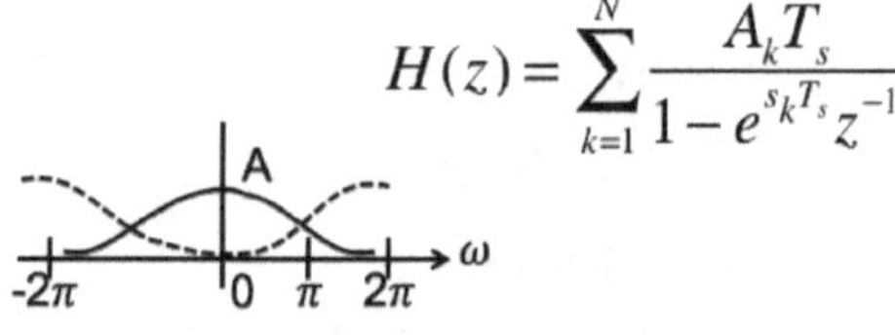

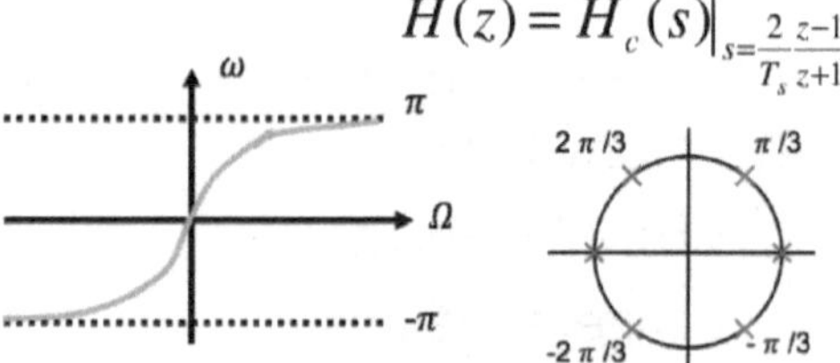

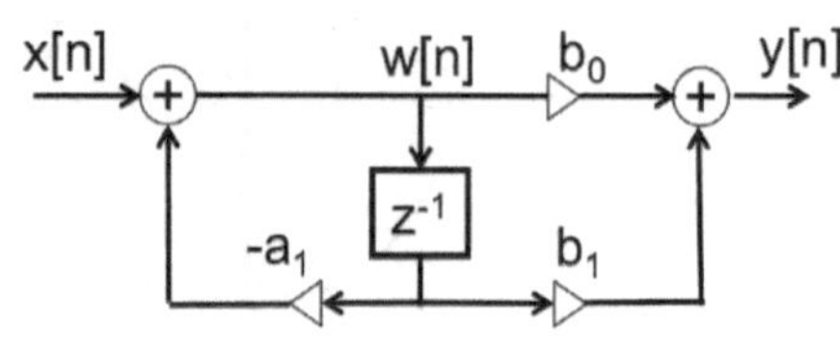

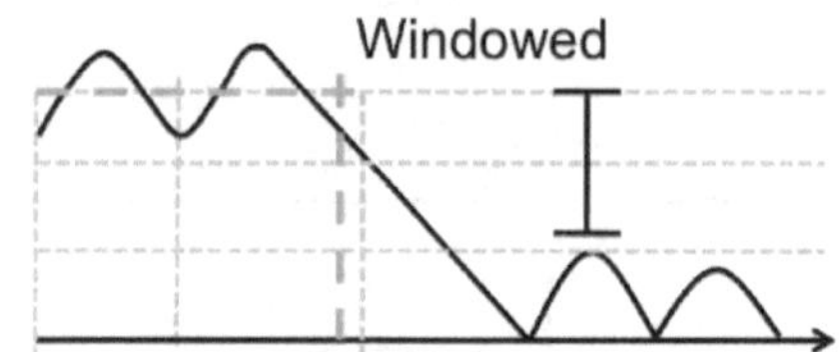

Basic Signals

Discrete-Time Signals

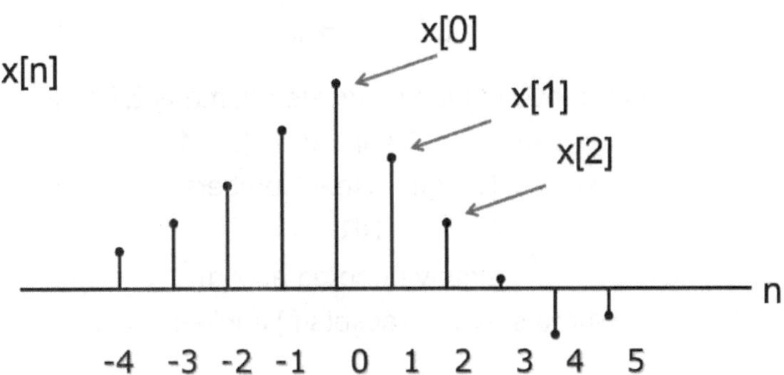

Discrete-Time Signal: $x[n]=x(nT_S)$

- In sampled system, $x[n]= x(n\ T_S)$

[ ] Square bracket denotes discrete-time signal

( ) parentheses denotes continuous-time signal

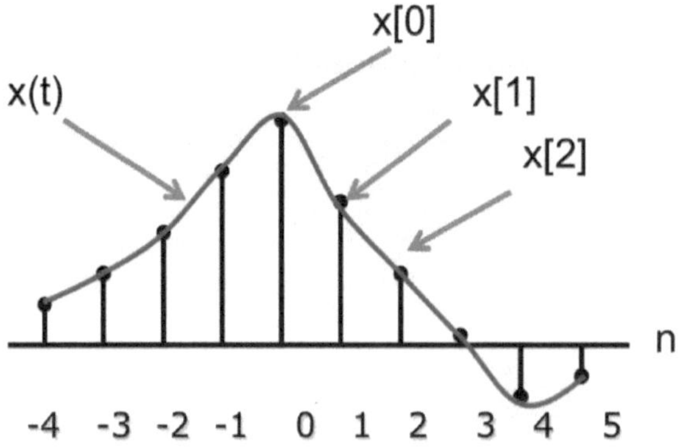

x[0]

x(t)

x[1]

x[2]

n

-4 -3 -2 -1 0 1 2 3 4 5

Sequences

What is a sequence?

An order set, order counts

Mathematically, a sequence is best seen as a function x:

$$x : \mathbb{Z} \rightarrow \mathbb{R}$$

where above notation indicates "x maps Z into R"

Domain = Z = Integers {….-1,0,1,…}

Range = Real Numbers =R

OR

simply denoted as x[n],

where square brackets [ ] implies $n \in Z$

Sequences

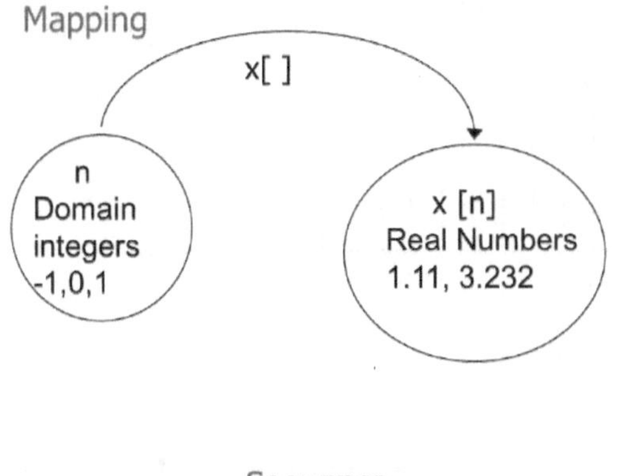

Table

n	x [n]
-2	2.2
-1	1.8
0	4
1	2.3
2	3.7

Sequence
(ordered set)

X= {x [n]} ={ ....x[-1], x[0], x[1], ... }

Important Discrete-Time Signals

- In sampling a continuous-time signal, $x[n]= x(n T_s)$
  - where x(t) is the continuous-time function,
  - and T_s is the sampling period

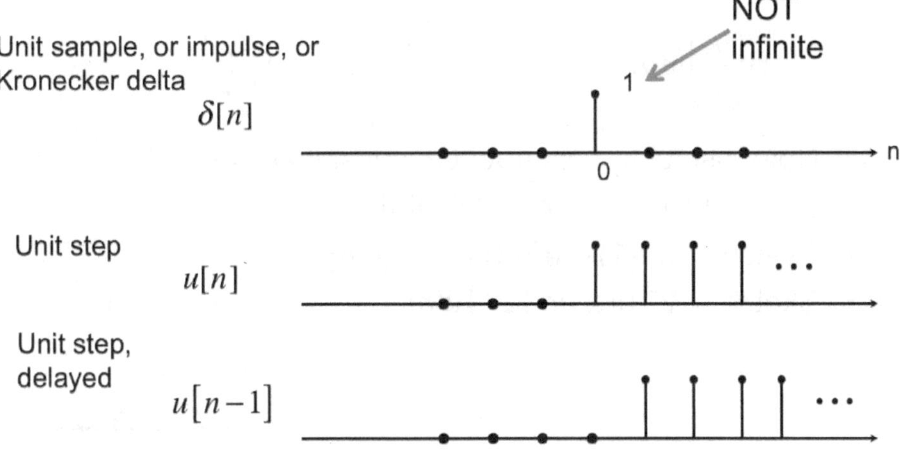

Important Discrete-Time Signals

Exponential sequence

$A\alpha^n$

A, α are real in plotted example

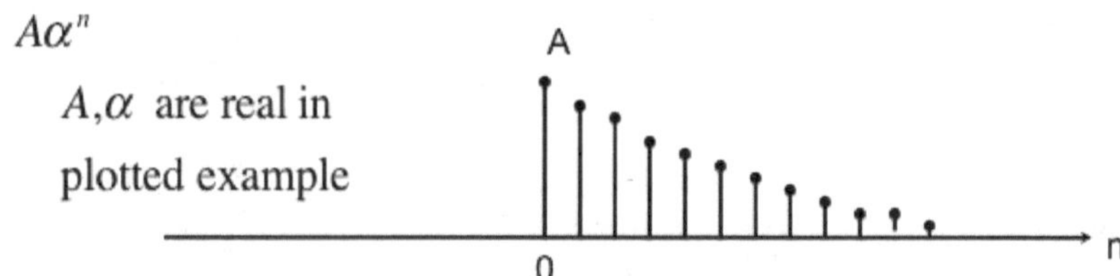

Sinusoid

$\cos[\omega_0 n + \varphi]$

$= \mathrm{Re}\left\{e^{j\omega_0 n} e^{j\varphi}\right\}$

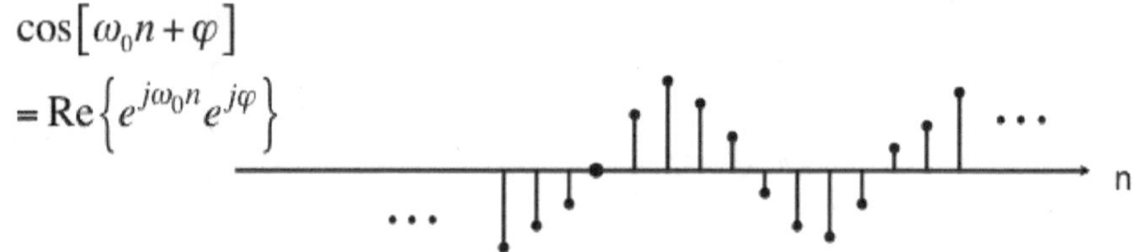

Pulse and Triangle Functions

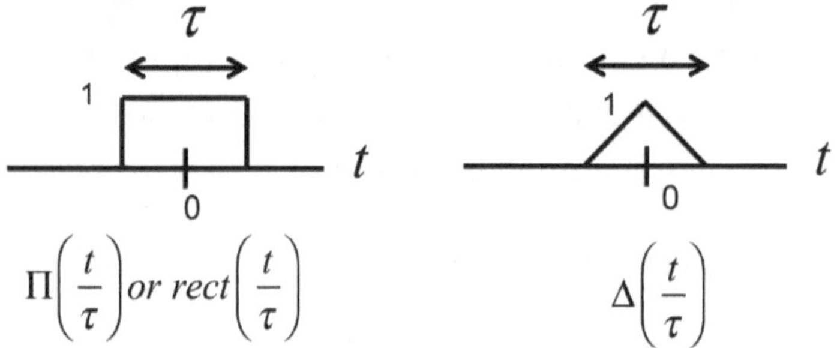

$\Pi\left(\frac{t}{\tau}\right) or\ rect\left(\frac{t}{\tau}\right)$ $\Delta\left(\frac{t}{\tau}\right)$

- Note: useful to construct casual signals, i.e.
  $e^{-at}u(t)$ or $e^{-an}u[n]$
- Sketch $u(t-1) - u(t-3)$ and $\Delta[n/4]$
- Sketch $2^{-t}u(t)$ and $2^{-n}u[n]$

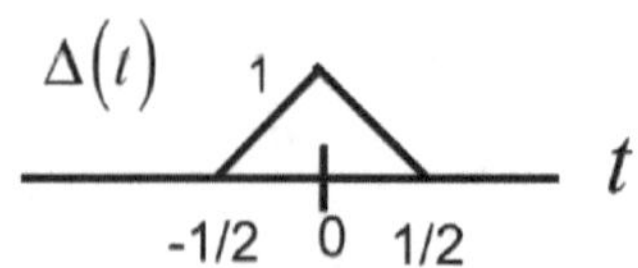

Exponential Sequences

- Exponential sequences

$$x[n] = A\alpha^n = |A|e^{j\varphi}|\alpha|^n e^{j\omega_0 n}$$

$$= |A||\alpha|^n e^{j(\omega_0 n+\varphi)} \qquad \alpha = |\alpha|e^{j\omega_0}$$

$$= |A||\alpha|^n \left(\cos(\omega_0 n+\varphi) + j\sin(\omega_0 n+\varphi)\right)$$

For A=1 and α=1, what is |x[n]| above?

- Euler identity

$$e^{j\theta} = \cos(\theta) + j\sin(\theta)$$

$$\cos(\theta) = \frac{1}{2}\left(e^{j\theta} + e^{-j\theta}\right)$$

$$\sin(\theta) = \frac{1}{2j}\left(e^{j\theta} - e^{-j\theta}\right)$$

Unit circle

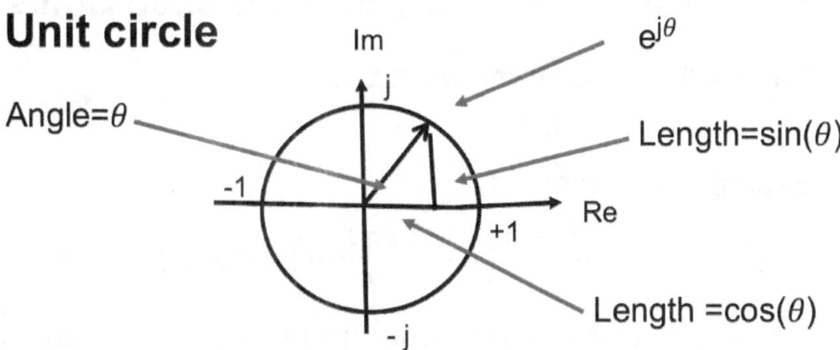

- $e^{j\theta}$= cos(θ) + j sin(θ) =α+jβ
- Can be viewed as the point in complex plane with coordinates (cos θ, sin θ)
- Can also be viewed as a vector, as shown above
- | $e^{j\theta}$ |=?

Unit Circle

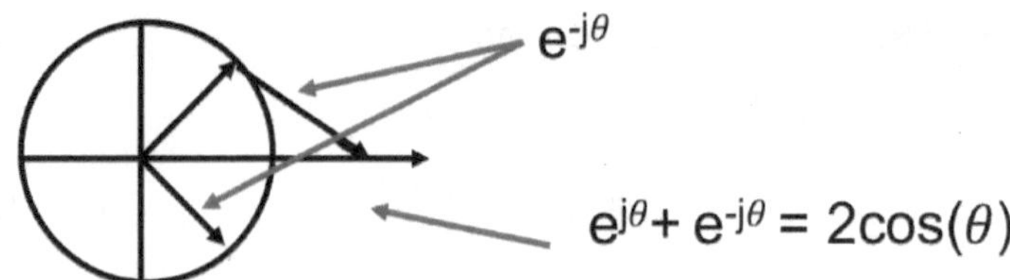

- As shown above, vector viewpoint is useful when adding
- Phasor notation can also be useful in polynomials
- Consider finding square root of j

  $(\gamma \angle \theta)^2 = \gamma^2 \angle 2\theta = j = 1\angle 90^{\circ}$

  or $(\gamma\, e^{j\theta})^2 = \gamma^2 e^{j2\theta} = j = 1\, e^{j\pi/2}$ (assuming γ is positive real)
- So, there are 2 roots to 2nd order polynomial

  $1\angle 45^{\circ}$ AND $1\angle 225^{\circ}$

Discrete-Time Frequency of a Sampled Sinusoid

$\omega_c = \Omega_c T_S$ discrete-time frequency

- Sampled with sampling period T_S
- $x[n] = x(t)|_{t=nT_S} = x(nT_S)$
- Example: $x(t) = \sin(\Omega_c t)$

$$x[n] = \sin\left(\frac{2\pi(nT_S)}{T_c}\right) = \sin\left(\frac{2\pi T_S}{T_c} n\right) = \sin(\Omega_c T_S n) = \sin(\omega_c n)$$

- Ω_c = continuous-time carrier frequency = $2\pi/T_c$ rad/s
- f_c = continuous-time carrier frequency = $1/T_c$ Hz
- Ω_S = Sampling frequency = $2\pi/T_S$
- f_S = sampling rate = $1/T_S$ Hz
- $\omega_c = \Omega_c T_S$ discrete-time carrier frequency, rad/sample

$$\omega_c = 2\pi\left(\Omega_c / \Omega_S\right) = \Omega_c T_S = 2\pi\left(f_c / f_S\right) = 2\pi\left(T_S / T_c\right)$$

Plot of Sampled Sinusoid

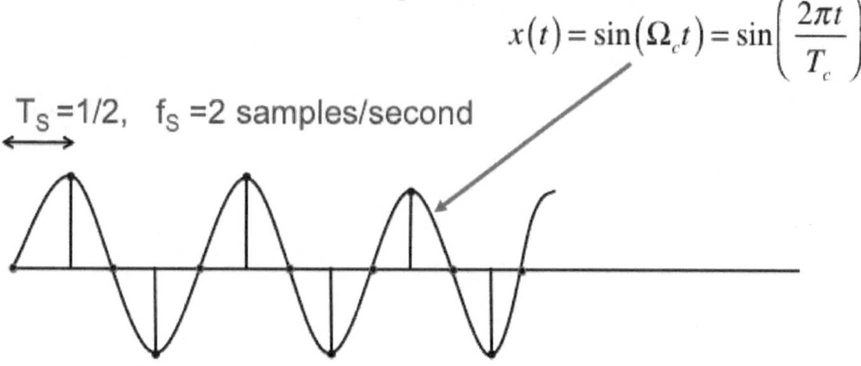

$T_c = 1/f_c = 2$

$x[n] = x(nT_S) = \sin(n\pi/2) = \{\ldots,0,1,0,-1,0,\ldots\}$

$\omega_c = 2\pi(f_c/f_S) = 2\pi(T_S/T_c) = \pi/2$ rad/sample

How many radians between each sample?

Aliasing

- Consider an analog sinusoid with extra 2π between samples
- Samples are exactly the same
- So, two different analog signals give rise to same samples
- To avoid aliasing, must have $f_s > 2 f_0$ (Nyquist)

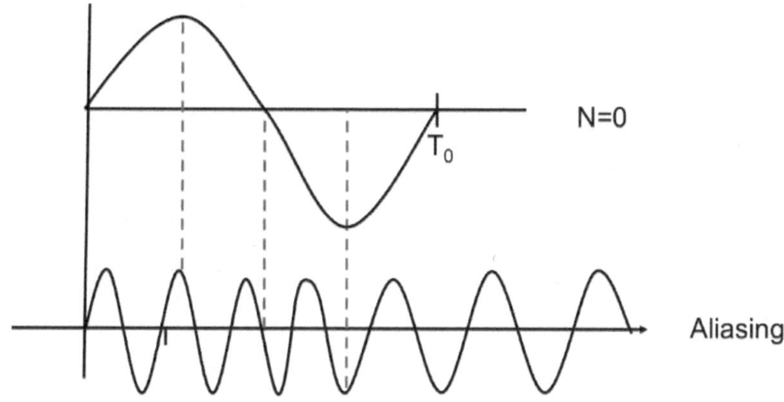

2 DISCRETE-TIME SIGNALS AND SYSTEMS

The lecture notes in this chapter cover discrete-time signals and systems, including linearity, time invariance, convolution, and BIBO stability.

Discrete-Time Signals and Systems

Discrete-time systems

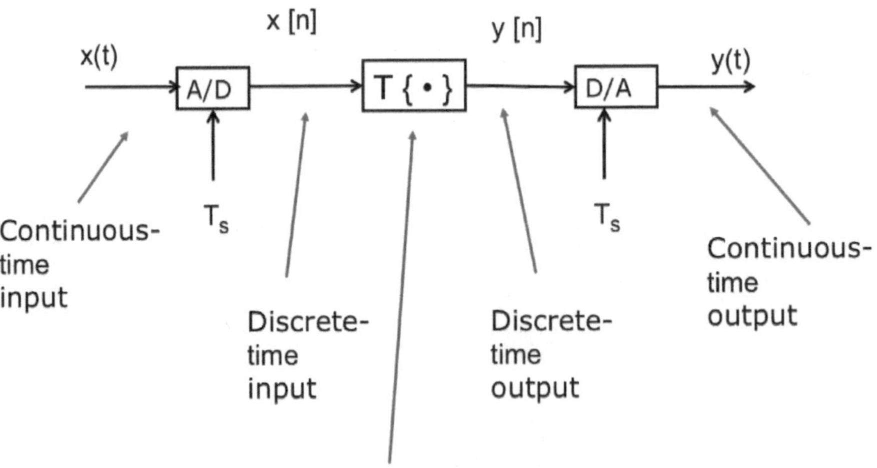

T { • } is the transformation operator denoting the DSP system or operation

T{ }

- y[n] = T { x [n] }
  where T { • } is transformation or operator or function relating discrete-time input to discrete-time output
- This is the most general form of input/output relationship
- Transformation/ Rule/ Mapping/ Function

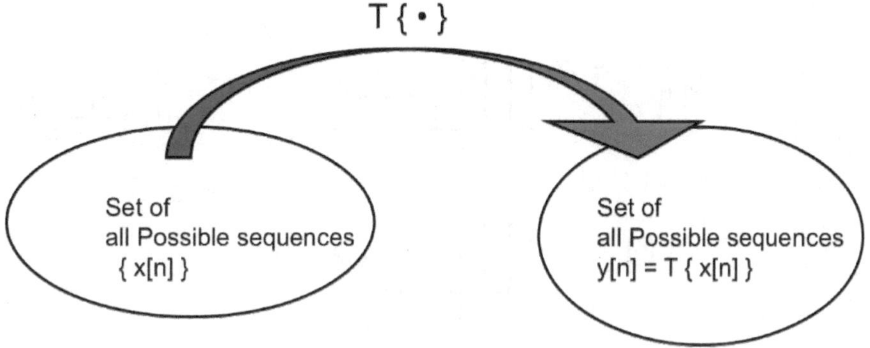

Example: Delay Systems

Delay of 1:   y[n] = x[n-1]

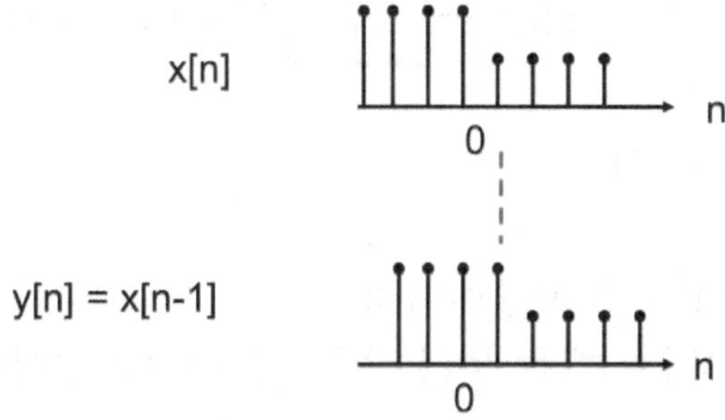

Example: Moving Average

2-point moving average: $y[n] = \dfrac{x[n] + x[n-1]}{2}$

- In words: today's average equals today's price plus yesterday's price, all divided by 2

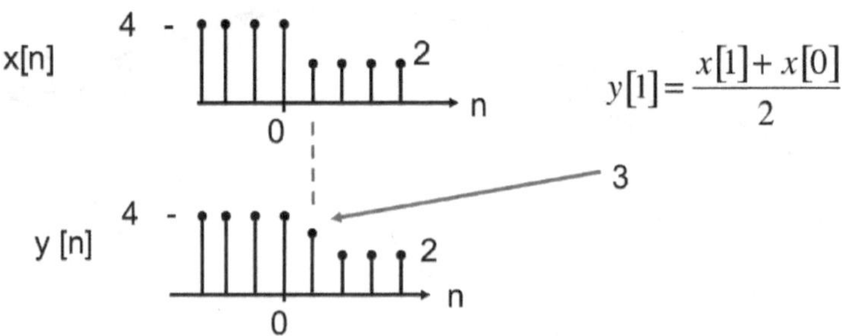

Discrete- Time System Properties

1. Memoryless system:

   y[n] only depends on x[n]

   So, output only depends on current input

2. Linear system:

   If x[n] → T{ } → y[n]= T{ x[n] }

   1) $y_1[n] = T\{ x_1[n] \}$

   2) and $y_2[n] = T\{ x_2[n] \}$

   then,

   the system is linear if and only if:

   3) $T\{ax_1[n] + bx_2[n]\} = aT\{x_1[n]\} + bT\{x_2[n]\} = ay_1[n] + by_2[n]$

   Envision 3 different tests in a laboratory

Properties

3. Time/Shift Invariant system:

   If $y[n]=T\{ x[n] \}$

   then, $y[n-n_d] = T\{ x[n-n_d] \} \; \forall \; n_d$ ← "for all"

   Informally, T{ } doesn't change over time

4. BIBO (Bounded Input Bounded Output) stability:

   If $|x[n]| \leq B_x \;\; \forall \; n$ (the input is bounded)

   then $|y[n]| \leq B_y \;\; \forall \; n$ (the output is bounded)

   where B_x and B_y are finite bounds.

   So, every bounded input produces bounded output

Example: accumulator

- Accumulator (running sum):

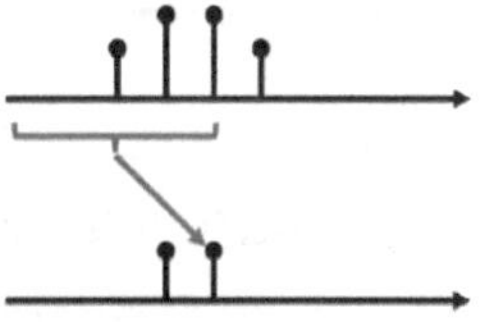

$$y[n] = \sum_{\alpha=-\infty}^{n} x[\alpha]$$

- What is y[n] when input x[n] is an impulse?
- What is y[n] when x[n] = u[n]?
- Is this system BIBO stable?
- What continuous-time system would have the same step response?

LTI (Linear Time Invariant) System

x[n] ⟶ T{ } ⟶ y[n]= T{ x[n] }

- Consider imposing linearity and time invariance on T{ }
- First. Break x[n] up into sum of many impulses times "weights" $x[n]=\sum_{k=-\infty}^{\infty} x[k]\delta[n-k]$
- Then, apply the transformation T{ }:

$$y[n]=T\{x[n]\}=T\left\{\sum_{k=-\infty}^{\infty} x[k]\delta[n-k]\right\}$$

$$=\sum_{k=-\infty}^{\infty} x[k]T\{\delta[n-k]\} \quad \textit{by linearity}$$

$$=\sum_{k=-\infty}^{\infty} x[k]h[n-k] \quad \textit{by time inv.}$$

$$\textit{result}: \quad y[n]=\sum_{k=-\infty}^{\infty} x[k]h[n-k]=x[n]*h[n] \quad \textit{convolution}$$

LTI System: Convolution

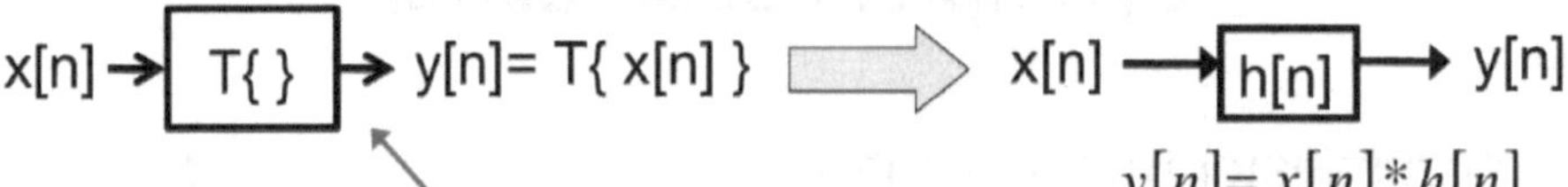

$$y[n]=x[n]*h[n]$$

- So:
  - Start with general system
  - Add 2 assumptions:
    - Linear
    - Time invariant
  - The result is convolution
- LTI is "common sense" behavior
  - Linear: knock on door twice as hard, the sound is twice as loud
  - Time invar.: knock on door tomorrow will make sound same as today
- Also, note:

$$h[n]*x[n]=\sum_{k=-\infty}^{\infty} h[k]x[n-k]=\sum_{k=-\infty}^{\infty} x[k]h[n-k]=x[n]*h[n]$$

Note continuous – time:

$$y(t)=x(t)*h(t)$$

$$=\int_{-\infty}^{\infty} x(\alpha)h(t-\alpha)d\alpha$$

A Brief Note About Functions, x[ ]

- Consider x[n] and x[1-n] shown below
- x[ ] embodies the function, *not* the argument inside "[ ]"
- When you put a "1" into x[ ], output is "9"
- When you put a "3" into x[ ], output is "5"
- So x[1-n] puts "1" into x[ ], at n=0, and is where a "9" is output

- So x[1-n] puts "3" into x[ ], at n= -2

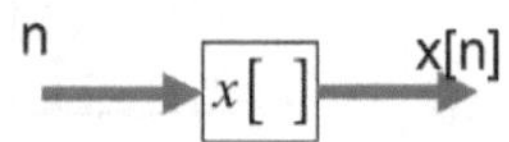

Function x[ ] as a machine
- Put "n" in
- Get x[n] out

$x[n]$: 9 at n = 1, 7, 5 at n = 3

$x[1-n]$: 5 at n = -2, 7, 9 at n = 0

What would plot be for 2x[1-n]?

Convolution: Sum of weighted- delayed h[n]'s

- Output y[n] = sum of weighted-delayed impulse responses

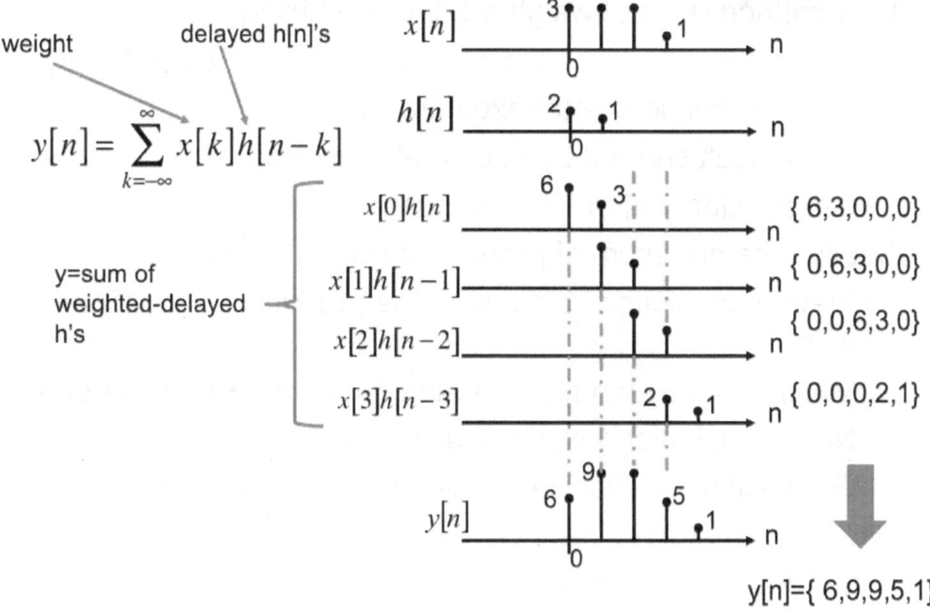

Convolution as Sum of Product of 2 Functions of k

- For each fixed n, view output y[n] as sum of the product of 2 functions of k

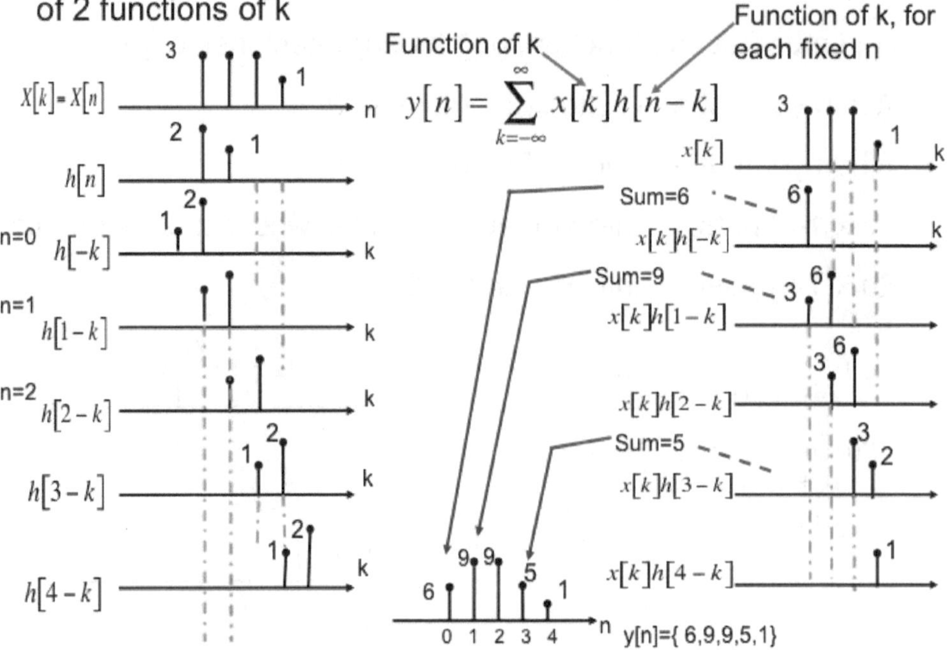

Compare Convolution Approaches

- Two views of convolution give same result
- First method (sum of weighted-delayed h[n]'s)
    - Views convolution as the sum of weighted, delayed, h[n]'s
    - Notice, plot horizontal-axes are "n"
    - Final result created by adding all functions
    - Get solution y[n] for all n "simultaneously"
- Second method (sum of product of two functions of k)
    - Views convolution as the sum the product of 2 functions of k
    - Multiply x[k] by h[n-k], then sum, for each fixed value of n
    - Notice each plot horizontal-axis is "k"
    - Get solution y[n] for each value n, "one at a time"
- Which is better to implement on computer?
- Which is better to implement in hardware?

Geometric Series Sum

Note:

A particularly useful formula for DSP

$$S_N = \sum_{k=0}^{N} \alpha^k = 1 + \alpha + \alpha^2 + ... + \alpha^N$$

$$S_N - \alpha S_N = \left(1 + \alpha + ... + \alpha^N\right) - \left(\alpha + \alpha^2 + ... + \alpha^{N+1}\right)$$

$$S_N(1-\alpha) = 1 - \alpha^{N+1}$$

Similarly: $$S_N = \sum_{k=0}^{N} \alpha^k = \frac{1-\alpha^{N+1}}{1-\alpha} \qquad N > 0$$

$$\sum_{k=N_1}^{N_2} \alpha^k = \frac{\alpha^{N_1} - \alpha^{N_2+1}}{1-\alpha}; \qquad N_2 > N_1$$

Convolution Example

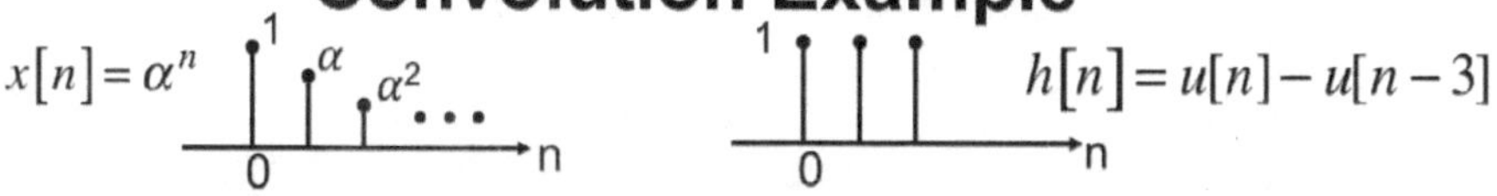

- Convolve the two functions above, using the "product of functions of k" method for convolution:

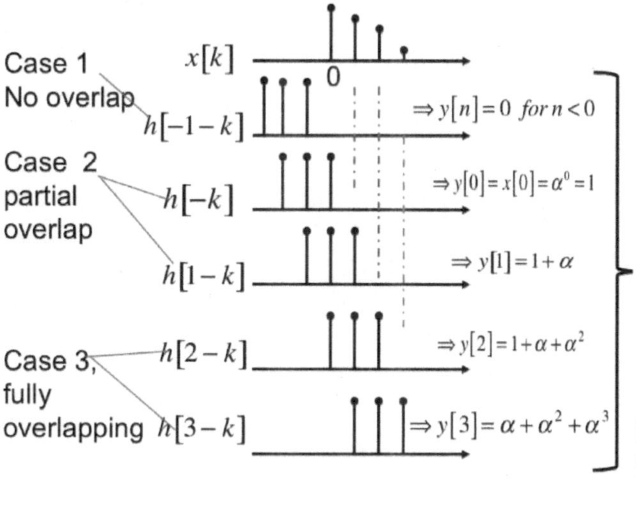

$$y[n] = \begin{cases} 0; & n < 0 \\ \sum_{k=0}^{n} \alpha^k; & 0 \le n < 2 \\ \frac{\alpha^{n-2} - \alpha^{n+1}}{1-\alpha}; & n \ge 2 \end{cases}$$

if α=0.5

y[n]={1,1.5,1.75,0.875,...}

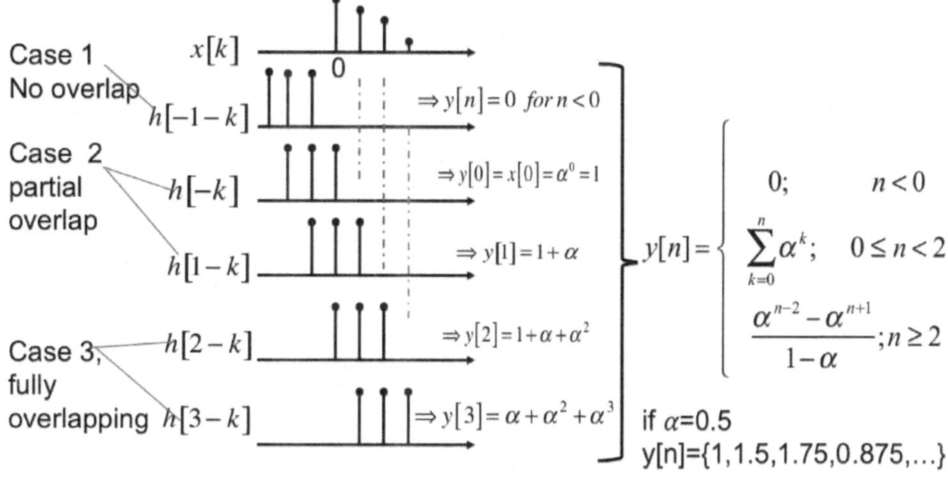

LTI System Summary

- For an LTI system, the output y[n] is the convolution of the input x[n] with the system impulse response h[n]:

$$\begin{aligned} y[n] &= x[n] * h[n] \\ &= h[n] * x[n] \end{aligned}$$

- Where convolution for a discrete-time system is defined as

$$y[n] = x[n] * h[n] = \sum_{k=-\infty}^{\infty} x[k]h[n-k]$$

LTI Properties

- Cascade systems:

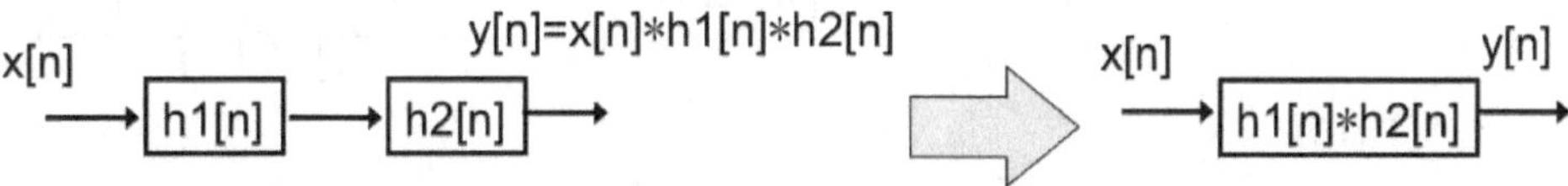

- Parallel systems:

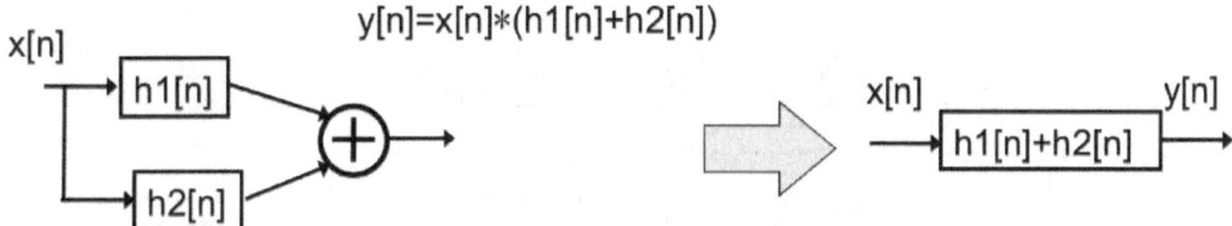

Cascade Proof

$$
\begin{aligned}
y[n] &= \left(x[n] * h_1[n]\right) * h_2[n] \\
&= \left(\sum_{k=-\infty}^{\infty} x[k]h[n-k]\right) * h_2[n] \\
&= \sum_{q=-\infty}^{\infty}\left(\sum_{k=-\infty}^{\infty} x[k]h_1[q-k]\right)h_2[n-q] \\
&= \sum_{k=-\infty}^{\infty} x[k]\sum_{q=-\infty}^{\infty} h_1[q-k]h_2[n-q] \\
&= \sum_{k=-\infty}^{\infty} x[k]\sum_{\alpha=-\infty}^{\infty} h_1[\alpha]h_2[(n-k)-\alpha] \\
&= x[n] * \left(h_1[n] * h_2[n]\right)
\end{aligned}
$$

BIBO Stability for LTI System

- Recall BIBO stability: bounded output for bounded input
- As before, the input is bounded by finite B_x:

$$|x[n]| \le B_x$$

x[n] → h[n] → y[n]

- Then:

$$y[n] = \sum_{k=-\infty}^{\infty} x[k]h[n-k] = \sum_{k=-\infty}^{\infty} h[k]x[n-k]$$

$$|y[n]| = \left| \sum_{k=-\infty}^{\infty} h[k]x[n-k] \right| \le \sum_{k=-\infty}^{\infty} |h[k]|\,|x[n-k]|$$

$$\le \sum_{k=-\infty}^{\infty} |h[k]|\,B_x \le B_x \sum_{k=-\infty}^{\infty} |h[k]|$$

$$\boxed{so, BIBO\ stable\ if \quad \sum_{k=-\infty}^{\infty} |h[k]| < \infty}$$

Example System

- Example: 3-point average

$$y[n] = \frac{x[n] + x[n-1] + x[n-2]}{3}$$

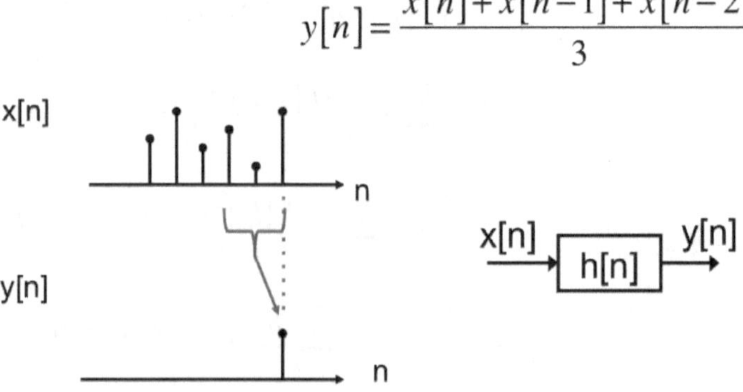

- What is impulse response of this system, h[n]?
- Is this system LTI?
- Is the system BIBO stable?

Example Cascade

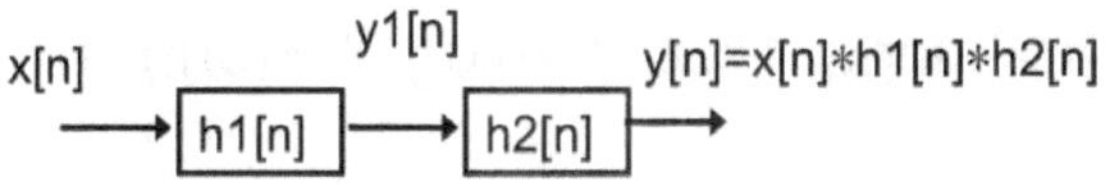

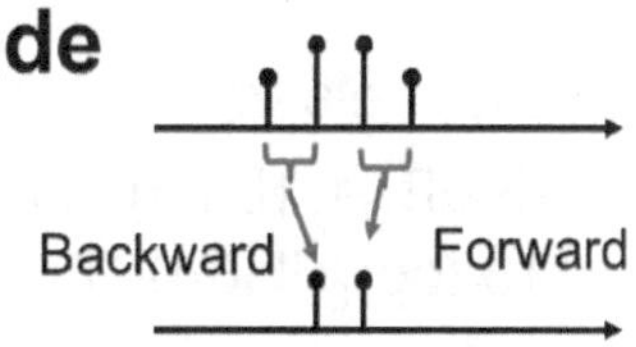

Definitions:

$$y_1[n] = x[n] - x[n-1] \quad \text{Backward Difference}$$

$$y[n] = y_1[n+1] - y_1[n] \quad \text{Forward Difference}$$

Impulse responses are:

$$h_1[n] = \delta[n] - \delta[n-1]$$

$$h_2[n] = \delta[n+1] - \delta[n]$$

Instead of using convolution, solve by direct substitution

$$y[n] = y_1[n+1] - y_1[n]$$

$$= (x[n+1] - x[n]) - (x[n] - x[n-1])$$

$$= x[n+1] - 2x[n] + x[n-1]$$

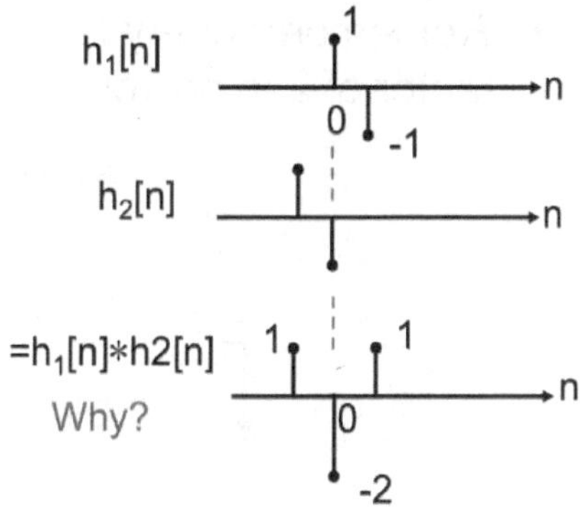

Does y[n] approximate a 2nd derivative?

Response to Complex Exponential

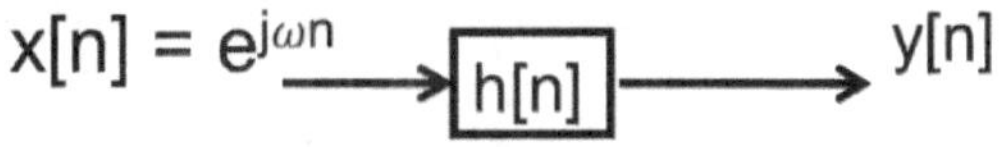

$$y[n] = h[n] * x[n] = \sum_{k=-\infty}^{\infty} h[k]x[n-k] = \sum_{k=-\infty}^{\infty} h[k]e^{j\omega(n-k)}$$

$$= e^{j\omega n} \sum_{k=-\infty}^{\infty} h[k]e^{-j\omega k} = e^{j\omega n} H(\omega)$$

- Frequency response $H(\omega)$
- Similarly for $\sin[\omega n]$, $\cos[\omega n]$
- Will return to this later: DTFT
- $H(\omega)$ arises naturally from convolution

Review of dB

- Decibels (dB):
- -3 dB = 10 $\log_{10}(1/2)$ = 20 $\log_{10}(1/\sqrt{2})$ = $20\log_{10}(|H(\omega)|)$ = 10 $\log_{10}(|H(\omega)|^2)$

  dB is always 10 $\log_{10}$(power ratio)

  10 $\log_{10}$(power ratio) = $10\log_{10}$((voltage ratio)^2)

  = 20 $\log_{10}$(voltage ratio)
- What is $|H(\omega)|$ at the 6dB bandwidth? ½
- Remember: every factor of 10 in power = 10 dB and every factor of 2 in power = 3 dB

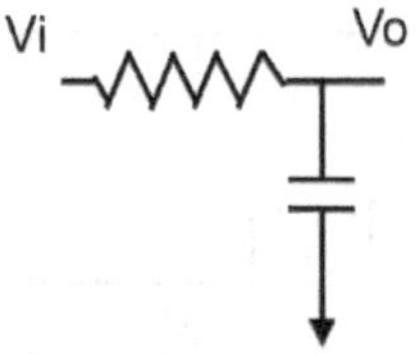

Vi Vo

$$H(s) = \frac{V_o(s)}{V_i(s)} = \frac{1/sC}{R + 1/sC} = \frac{1}{1 + sRC}$$

$$H(\Omega) = \frac{1}{1 + j\Omega RC}$$

$$= \frac{1}{1+j}; \quad at\ \Omega = \frac{1}{RC}$$

For RC filter:

What is 3 dB bandwidth?

What is half-power bandwidth?

3 SAMPLING AND DIFFERENCE EQUATIONS

The lecture notes in this chapter cover the topics of sampling, aliasing, Nyquist rate, quantization noise, and difference equations.

Sampling and Difference Equations

A/D & D/A Conversion, x(t) and x[n]

- Consider the ADC/DAC system below (simple A-to-D-to-A system)
- This system embodies the fundamental principals of digitizing signals
- ADC: Analog-to-Digital Converter, DAC: Digital-to-Analog Converter
- As illustrated, let $x[n]=x(nT_S)$ be the discrete-time signal

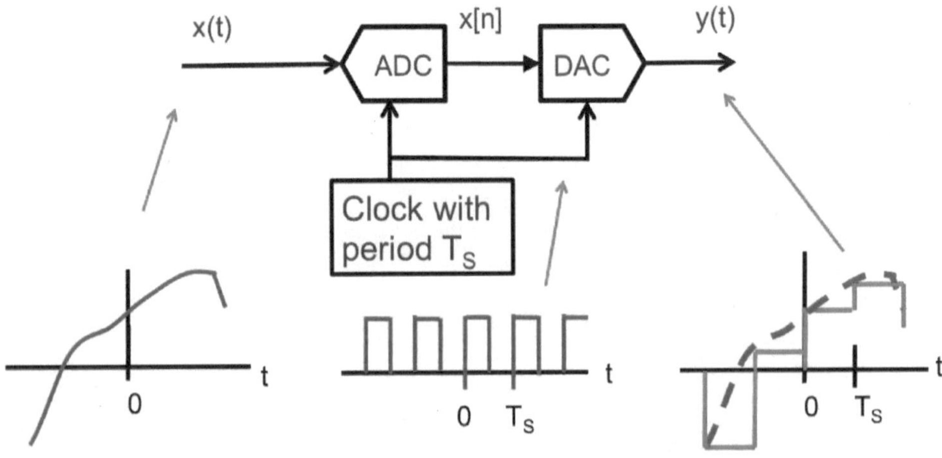

Sampling (Model of A/D+D/A Conversion)

- Consider ADC model below, where h(t) = $\Pi((t-T_s/2) / T_s)$
- $\Pi(t/\tau)$ is rectangular pulse centered at zero, width τ and height 1
- Below, $x_s(t)$ is the sampled signal
- This model has the same input/output as the ADC+DAC system

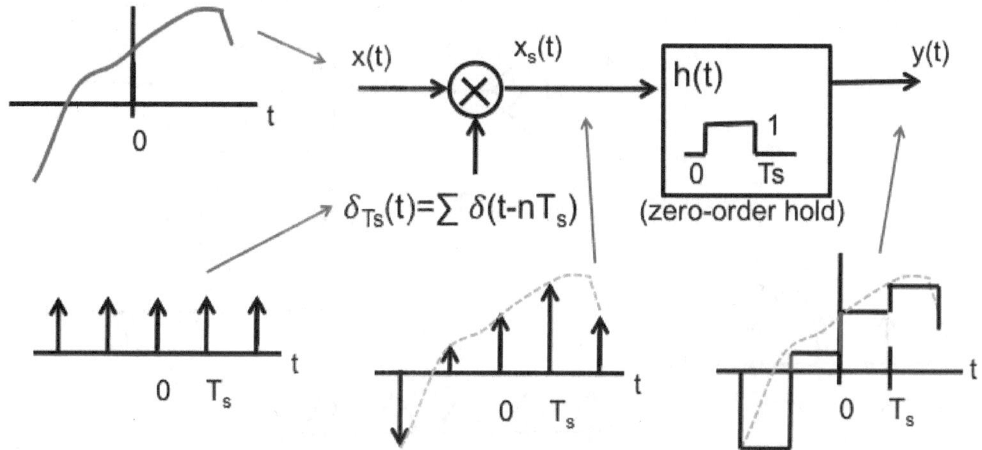

Sampling

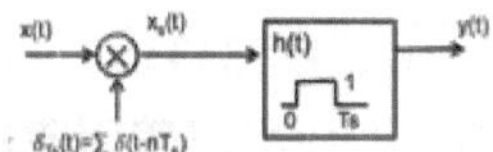

$$\Pi(t/T_s) \Leftrightarrow T_s\,\mathrm{sinc}(\Omega T_s/2)$$

Take the Fourier transform at each point

$$g_1(t)g_2(t) \Leftrightarrow \frac{1}{2\pi}G_1(\Omega)*G_2(\Omega)$$

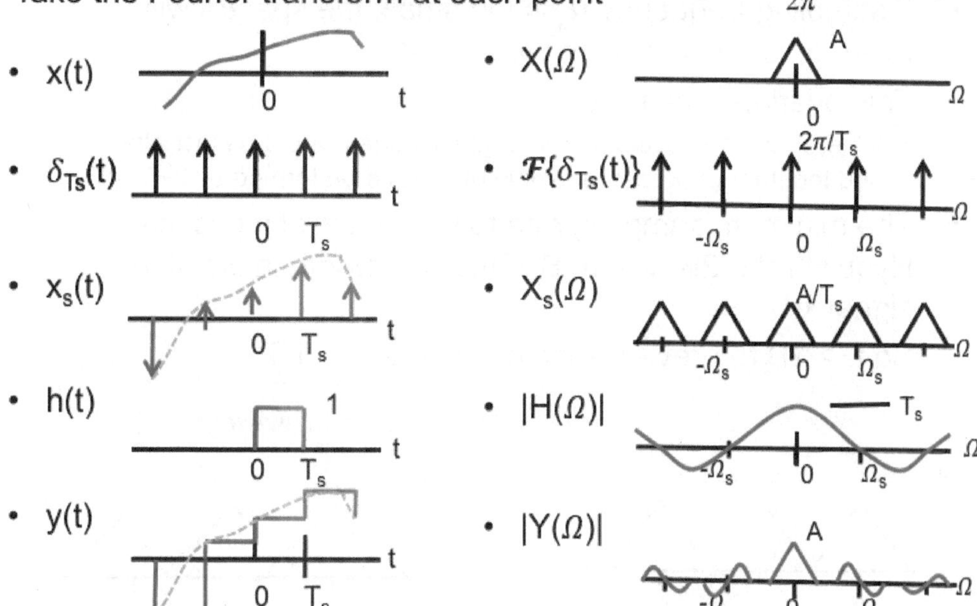

Sampling Redrawn Using Spectra

- This model has the same input/output as the ADC+DAC system
- h(t)=Π\{(t-Ts/2)/(Ts)\},

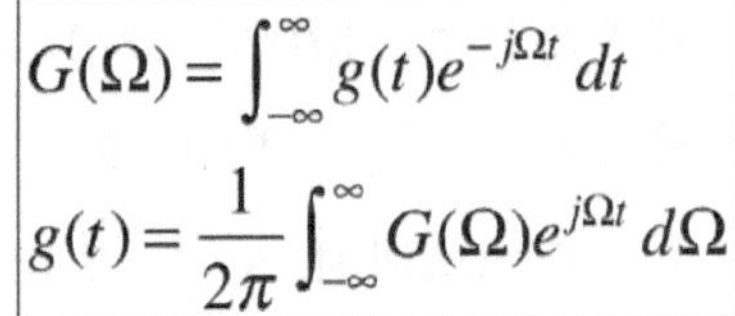

$$G(\Omega) = \int_{-\infty}^{\infty} g(t) e^{-j\Omega t}\, dt$$

$$g(t) = \frac{1}{2\pi} \int_{-\infty}^{\infty} G(\Omega) e^{j\Omega t}\, d\Omega$$

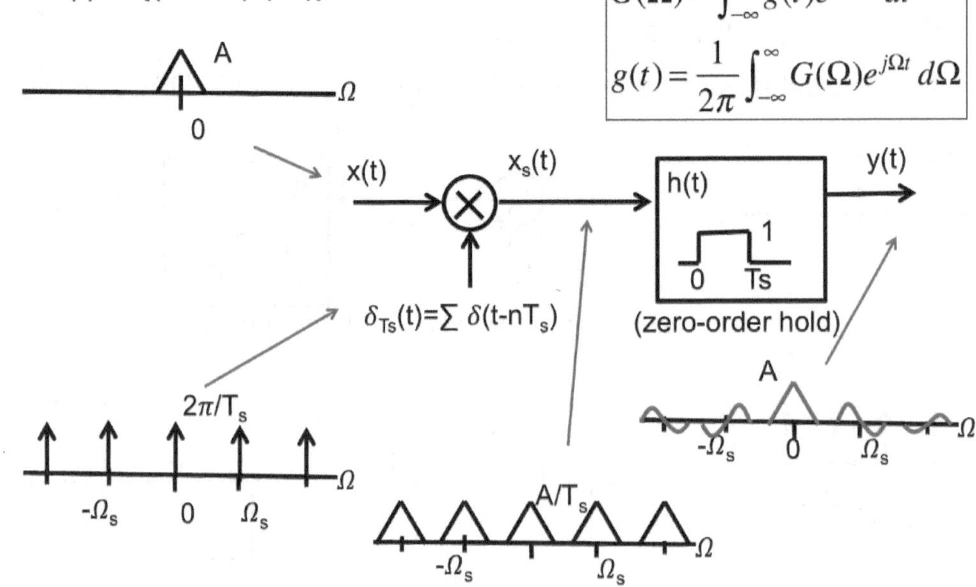

Nyquist Rate and Aliasing

- Consider the spectrum of the sampled signal $X_s(\Omega)$ below
- If sampling frequency Ω_s is too small, the spectra will overlap, and information will be lost
- This overlap is aliasing
  - Where overlap occurs below, original values would be lost after being add together (a measured value of 4 could be formed by 2+2 or 3+1)
- The minimum sampling rate to prevent aliasing is the Nyquist rate, 2B, where B is the bandwidth in Hz of the signal x(t)
- So $f_s > 2B$ to prevent aliasing, where $f_s = 1/T_s$

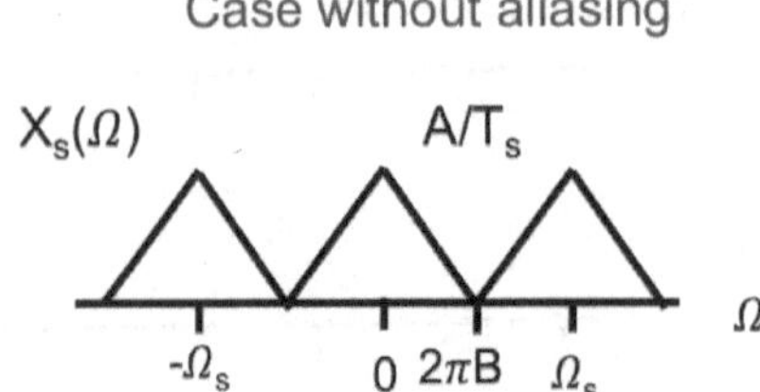

aliasing

Case with aliasing

$X_s(\Omega)$

$-\Omega_s$ 0 Ω_s Ω

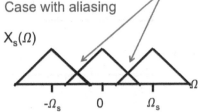

Perfect Reconstruction Filter

- The filter that modeled the ADC-DAC pair had h(t) = Π((t-T_s/2) / T_s), and is referred to as a zero-order hold filter
- This filter had the effect of distorting the time domain signal by changing the smooth x(t) into a "staircase/stepwise" approximation y(t)
- In the frequency domain, H(Ω) multiplied the signal spectrum by a sinc() function, and did not remove the harmonics from the spectrum
- Below, if h(t) is an ideal lowpass filter with H(Ω) =T_s $\Pi(\Omega/\Omega_s)$, then the output of the system will exactly equal the original input; y(t) = x(t) and Y(Ω) = X(Ω)

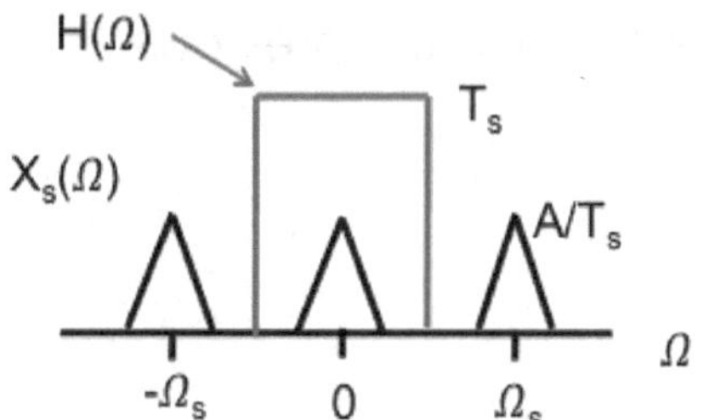

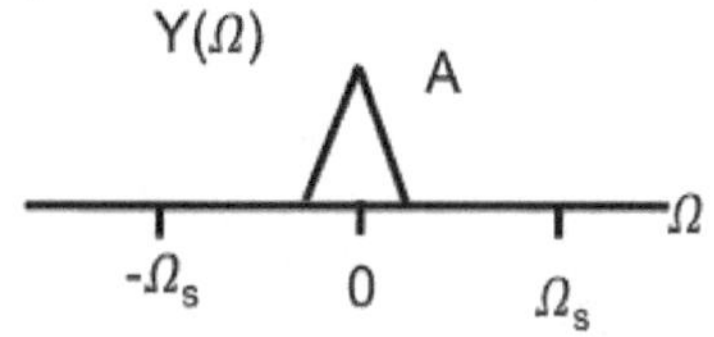

Perfect Reconstruction System

- Consider model below, where H(Ω) =T_s $\Pi(\Omega/\Omega_s)$
- Now the output signal y(t) exactly equals the input signal x(t)

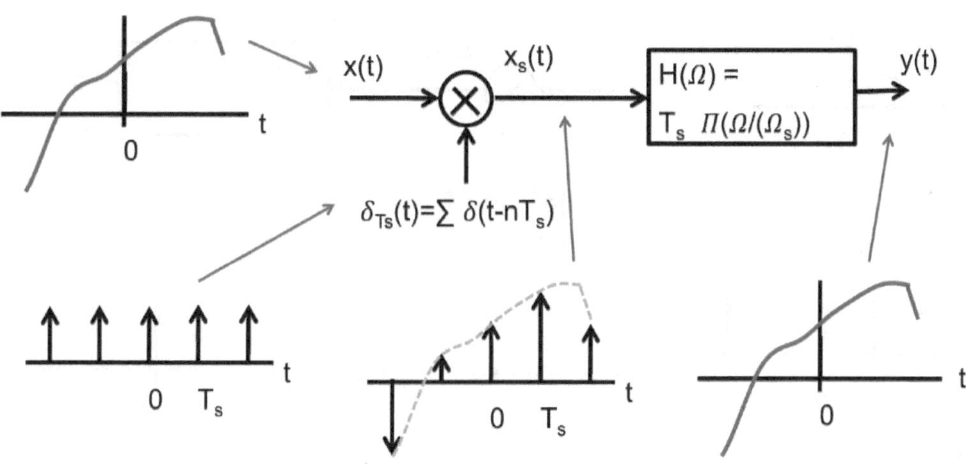

Sampling Theorem

- Nyquist-Shannon Sampling theorem
  - If a signal is *strictly bandlimited* to B Hz, then it can be *perfectly reconstructed* from its samples, if the signal is sampled at sampling rate greater than 2B Hz.
- This means that all time points between the sample points can be exactly recovered
- This theorem is essentially the same result as the perfect reconstruction systems given in the previous slides, since y(t) = x(t) !!

Perfect Reconstruction (Convolution View)

- The impulse response of the perfect reconstruction filter is $h(t)=\text{sinc}(\pi t/T_s)$
- When convolved with $x_s(t)$, the sinc() functions interpolate between samples as illustrated below

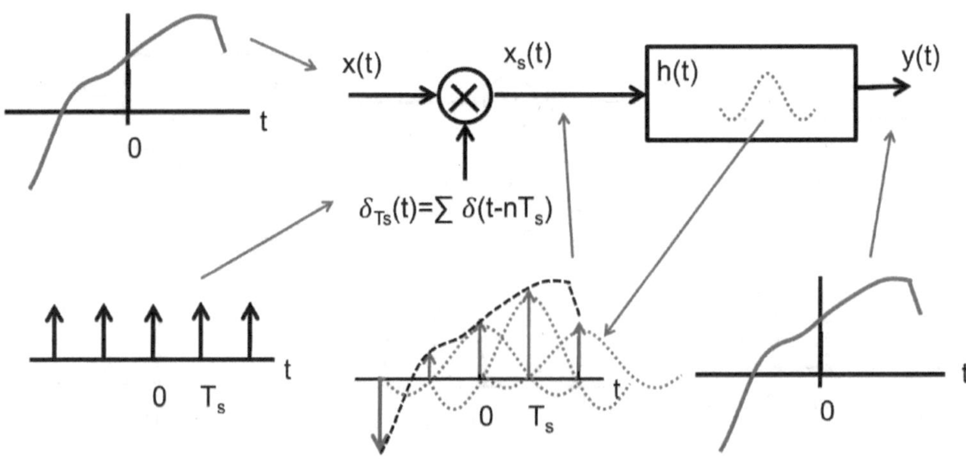

ADC and DAC RMS Quantization Noise

- The effect of random quantization noise error q(t) is a noise voltage added to the analog signal being digitized
- This noise is in both the ADC and the DAC
- The noise q(t) has a uniform pdf from -Δv/2 to Δv/2, where Δv is the size of the voltage steps/increments of the ADC or DAC
- Mean noise $\mu_q = 0$, and rms noise voltage $\sigma_q = \Delta v/\sqrt{12}$
- Also, maximum full-bandwidth signal-to-noise in ADC ≈1.8 + 6N dB, where N is number of bits

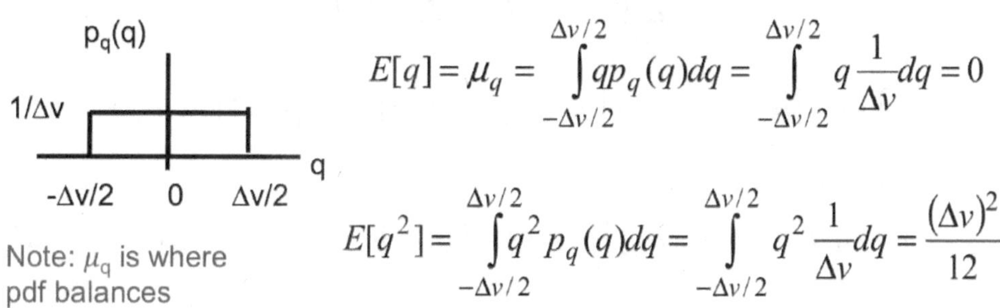

$$E[q] = \mu_q = \int_{-\Delta v/2}^{\Delta v/2} q p_q(q) dq = \int_{-\Delta v/2}^{\Delta v/2} q \frac{1}{\Delta v} dq = 0$$

$$E[q^2] = \int_{-\Delta v/2}^{\Delta v/2} q^2 p_q(q) dq = \int_{-\Delta v/2}^{\Delta v/2} q^2 \frac{1}{\Delta v} dq = \frac{(\Delta v)^2}{12}$$

Bandwidth

- Recall definitions of bandwidth

- Ideal low-pass systems
  - See figure at right
  - Bandwidth is 0 to cutoff

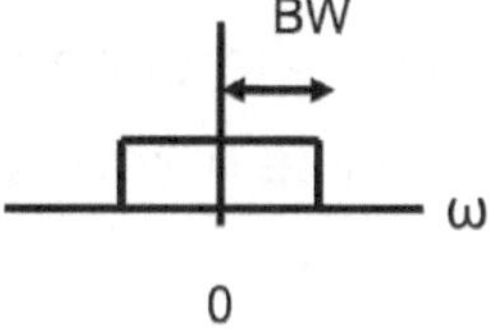

- Ideal bandpass systems
  - See figure at right
  - Bandwidth is full width of positive frequencies

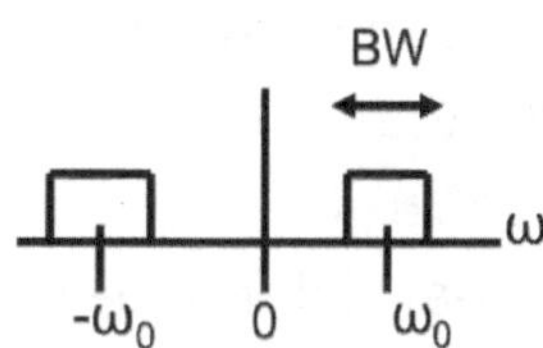

Difference Equations

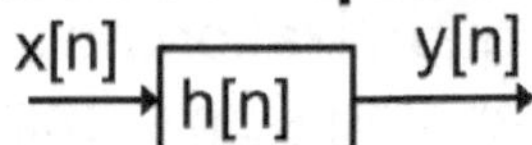

- Difference equations
  - Provide another way to represent discrete-time systems
  - Difference equations are not unique; a given system may be defined by a variety of difference equations
  - Difference equations are analogous to differential equations used in continuous-time systems

$$\sum_{p=0}^{N} a_p y[n-p] = \sum_{q=0}^{M} b_q x[n-q]$$

$$a_0 y[n] + a_1 y[n-1] + ..a_N y[n-N] = b_0 x[n] + b_1 x[n-1] + ..b_M x[n-M]$$

- In general, let $a_0=1$
  $y[n] + a_1 y[n-1] + ..a_N y[n-N] = b_0 x[n] + b_1 x[n-1] + ..b_M x[n-M]$

Difference Equations

- Recall the accumulator definition:

$$y[n] = \sum_{k=-\infty}^{n} x[k]$$

- This is in difference equation format
- The impulse response is

$$h[n] = \sum_{k=-\infty}^{n} x[k] \Big|_{x[n]=\delta[n]} = \sum_{k=-\infty}^{n} \delta[k] = u[n]$$

- The form of the difference equation is not unique (is this a good property or a problem?)

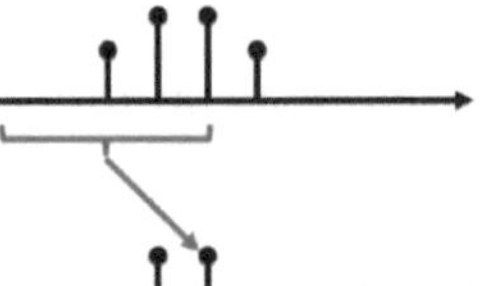

- Another difference equation for the system is:

$$y[n] = \sum_{k=-\infty}^{n} x[k] = x[n] + \sum_{k=-\infty}^{n-1} x[k] = x[n] + y[n-1]$$

Block diagram based on difference eqn.

$$y[n] - y[n-1] = x[n]$$

Difference Equations

- Another representation could be:

$$y[n]=\sum_{k=-\infty}^{n-2} x[k]+x[n-1]+x[n]$$

$$y[n]=y[n-2]+x[n-1]+x[n]$$

$$y[n]-y[n-2]=x[n]+x[n-1]$$

- And another representation:

$$y[n]=y[n-2]+x[n]+x[n-1]$$

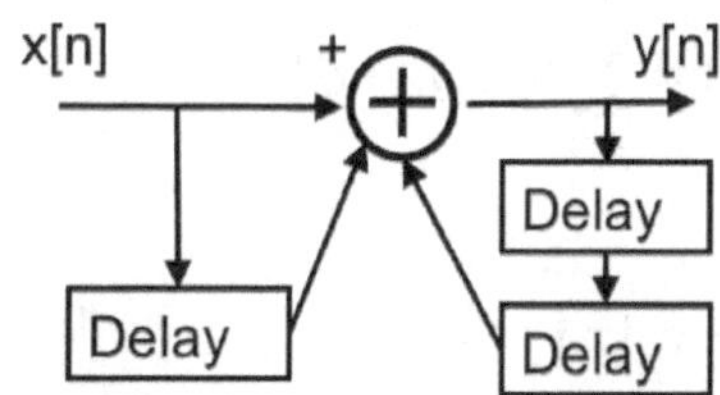

Block diagram based on difference eqn.

***Note:* all of these difference equation forms have the same impulse response h[n]**

Since there are many possible solutions, the choice becomes "the art of engineering"

***Note:* each difference equation leads to a different block diagram**

Using the block diagram, what is h[n]?
Is h[n] the same as before?
Does this require more hardware?

Example

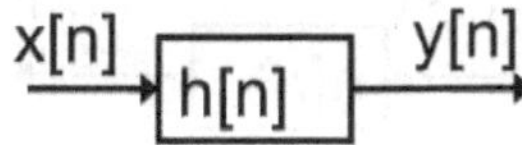

- Convolve the following two functions to find the system output, y[n] :

  x[n]= {1,2,3,4} (ramp)

  h [n]= {1,-1} (backward difference)

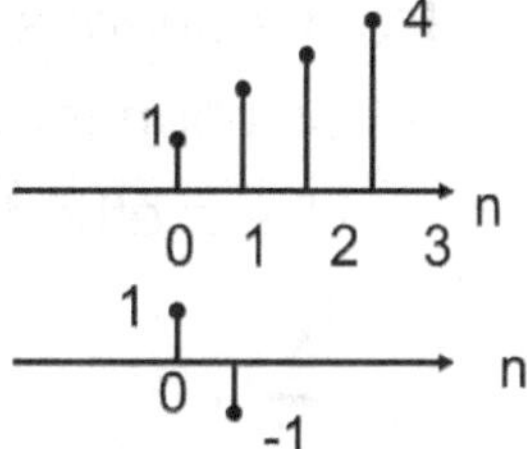

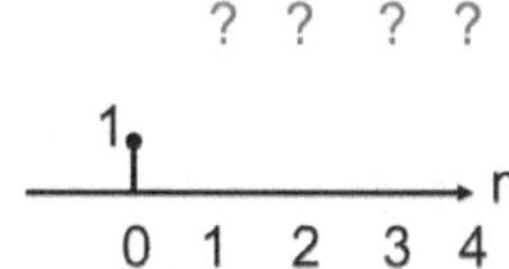

Also note the difference equation for backward difference:

$$y[n]=x[n]-x[n-1]$$

Solving Difference Equations

x[n] → h[n] → y[n]

- Difference equations
  - Solution is analogous to differential equations
    1. Find homogeneous (transient) solution $y_h[n]$
    2. Find particular (forced) solution $y_p[n]$
    3. Use initial conditions to find solution y[n]

$$\sum_{p=0}^{N} a_p y[n-p] = \sum_{q=0}^{M} b_q x[n-q]$$
$$solution:$$
$$y[n] = y_h[n] + y_p[n]$$

Homogeneous solution

x[n] → h[n] → y[n]

- Homogeneous (transient) solution $y_h[n]$
  - Solution for no input, x[n]=0
  - Use method of guessing: guess $y[n]=A\alpha^n$

$$\sum_{p=0}^{N} a_p y_h[n-p] = \sum_{q=0}^{M} b_q x[n-q] = 0$$
$$guess: y_h[n] = A\alpha^n$$
$$\sum_{p=0}^{N} a_p A\alpha^{n-p} = 0$$
$$A\left(a_0\alpha^n + a_1\alpha^{n-1} + \cdots a_N\alpha^{n-N}\right) = A\alpha^{n-N}\left(a_0\alpha^N + a_1\alpha^{N-1} + \cdots a_N\right) = 0$$
$$solve\ for\ N\ roots, \alpha_p, of\ polynomial\ a_0\alpha^N + a_1\alpha^{N-1} + \cdots a_N$$
$$then: y_h[n] = A_1\alpha_1^n + A_2\alpha_2^n + \cdots A_N\alpha_N^n \quad \text{(assuming simple roots)}$$

Particular solution

x[n] → h[n] → y[n]

- Particular (forced) solution $y_p[n]$
  - Solution with input, $x[n] \neq 0$
  - Use method of guessing: guess y[n] has same form as x[n]
  - For example, if $x[n]=B\beta^n$ guess that $y[n]=G\,\gamma^n$

$$\sum_{p=0}^{N} a_p y_p\left[n-p\right]=\sum_{q=0}^{M} b_q x\left[n-q\right]$$
$$if\ x[n]=B\beta^n$$
$$guess: y_p[n]=G\gamma^n$$
$$G\left(a_0\gamma^n+a_1\gamma^{n-1}+\cdots a_N\gamma^{n-N}\right)=B\left(b_0\beta^n+b_1\beta^{n-1}+\cdots b_M\beta^{n-M}\right)$$
$$solve\ for\ G\ and\ \gamma$$
$$then: y_p[n]=G\gamma^n$$

Complete Solution

x[n] → h[n] → y[n]

- Complete solution y[n]
  - o $y[n] = y_h[n] + y_p[n]$
  - o Use initial conditions to solve for remaining constants

Example Solution of Difference Equation

x[n] h[n] y[n]

- Difference equation example:

$$\sum_{p=0}^{N} a_p y[n-p] = \sum_{q=0}^{M} b_q x[n-q]$$

$$y[n] - ay[n-1] = x[n]$$

$$let \;\; a = 0.5$$

$$so \;\;\; y[n] - 0.5y[n-1] = x[n]$$

$$initial\ conditions: y[0] = 2$$

$$input: \;\; x[n] = B\beta^n = 5(1/3)^n$$

$$solution:$$

$$y[n] = y_h[n] + y_p[n]$$

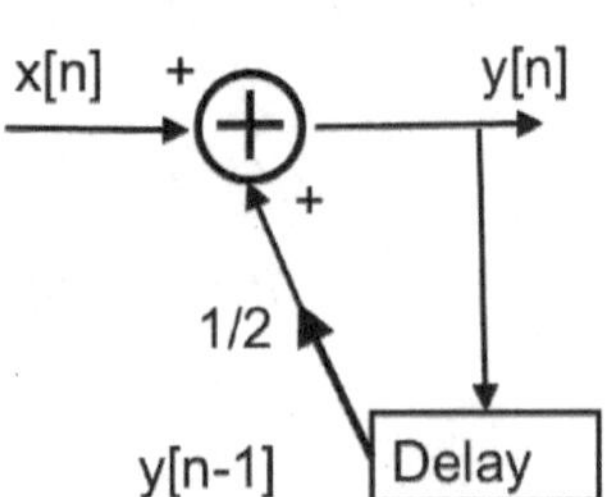

Example, continued

- Homogeneous (transient) solution $y_h[n]$
  - Use method of guessing: guess $y[n]=A\alpha^n$

$$y_h[n] - 0.5y_h[n-1] = x[n] = 0$$

$$guess: y_h[n] = A\alpha^n$$

$$A\alpha^n - 0.5A\alpha^{n-1} = A\alpha^{n-1}\left(\alpha^1 - 0.5\right) = 0$$

$$solve\ for\ 1\ root, \alpha_1, of\ polynomial; \quad \alpha_1 = 0.5$$

$$then: y_h[n] = A_1(0.5)^n$$

Example, continued

- Particular (forced) solution $y_p[n]$
  - Use method of guessing: guess y[n] has same form as x[n]

$$y_p[n]-0.5y_p[n-1]=x[n]$$
$$x[n]=B\beta^n$$

$$guess: y_p[n]=G\gamma^n$$
$$G\left(\gamma^n-0.5\gamma^{n-1}\right)=G\gamma^n\left(1-0.5\gamma^{-1}\right)=B\beta^n$$

$$\begin{aligned}&solve\ for\ G\ and\ \gamma;\\&\quad by\ comparison\ of\ both\ sides\ \gamma=\beta\\&\quad then\ G=B/\left(1-0.5\gamma^{-1}\right)\end{aligned}$$

$$then: y_p[n]=G\gamma^n=\frac{B\beta^n}{1-0.5\beta^{-1}}=\frac{5(1/3)^n}{1-1.5}=-10(1/3)^n$$

Example, continued

- Complete solution y[n]
  - o $y[n] = y_h[n] + y_p[n]$
  - o Use initial conditions to solve for remaining constants

$$y[n]=y_h[n]+y_p[n]=A_1(0.5)^n-10(1/3)^n$$
$$initial\ conditions: y[0]=2$$
$$y[0]=2=A_1(0.5)^0-10(1/3)^0=A_1-10$$
$$so\quad A_1=12$$
$$finally\ the\ complete\ solution:$$
$$y[n]=12(0.5)^n-10(1/3)^n$$

If a system has a nonzero transient response, is it LTI?

Recursive Diff Eq. Solution and IIR Systems

- Recursive solution:

$$\text{Example:} \quad y[n]-ay[n-1]=x[n]$$

$$\text{or} \quad y[n]=ay[n-1]+x[n]$$

assume system at rest and impulse input:

$$y[0]=ay[-1]+x[0]=0+1=1$$

$$y[1]=ay[0]+x[1]=a+0=a$$

$$y[2]=ay[1]+x[2]=a^2+0=a^2$$

$$y[3]=ay[2]+x[3]=a^3+0=a^3$$

$$\cdots so, \quad y[n]=a^n u[n]$$

Block diagram based on difference eqn.

•This is an infinite impulse response (IIR) system (infinite duration h[n])

•A finite impulse response (FIR) system would have a finite duration h[n]

FIR Systems

- FIR example (3-point sum):

$$\text{Example:} \quad y[n]=x[n]+x[n-1]+x[n-2]$$

assume system at rest and impulse input:

$$y[0]=x[0]+x[-1]+x[-2]=1+0+0=1$$

$$y[1]=x[1]+x[0]+x[-1]=0+1+0=1$$

$$y[2]=x[2]+x[1]+x[0]=0+0+1=1$$

$$y[3]=x[3]+x[2]+x[1]=0+0+0=0$$

$$\cdots so$$

$$y[n]=\delta[n]+\delta[n-1]+\delta[n-2]$$

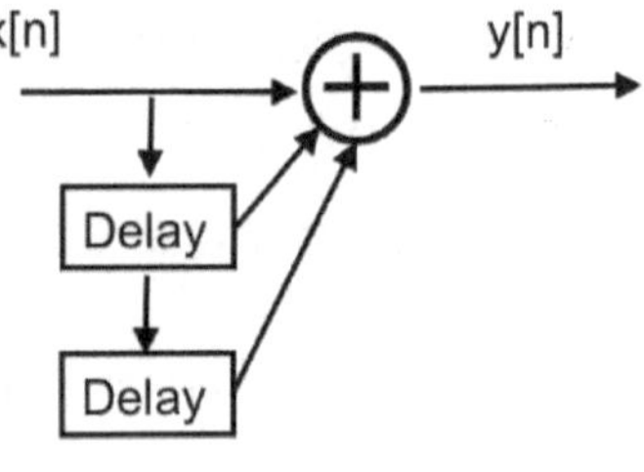

Block diagram based on difference eqn.

•This is a finite impulse response (FIR) system with a finite duration h[n]

Does this FIR system block diagram have feedback from the output?

4 DISCRETE-TIME FOURIER TRANSFORM

The lecture notes in this chapter present the DTFT (discrete-time Fourier transform) and frequency response of systems.

Discrete-Time Fourier Transform

DTFT

- Recall: $x_s(t) = x(t)\delta_{Ts}(t) = x(t) \sum \delta(t-nT_s)$
- So:

$$X_s(\Omega) = \int_{-\infty}^{\infty} x_s(t) e^{-j\Omega t}\, dt$$

$$= \int_{-\infty}^{\infty} \left(\sum_{n=-\infty}^{\infty} x(nT_s)\delta(t-nT_s) \right) e^{-j\Omega t}\, dt$$

$$= \sum_{n=-\infty}^{\infty} x(nT_s) \int_{-\infty}^{\infty} \delta(t-nT_s) e^{-j\Omega t}\, dt$$

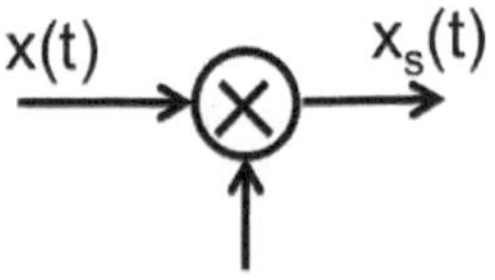

x(t)

$x_s(t)$

$\delta_{Ts}(t)=\sum \delta(t-nT_s)$

$$X_s(\Omega) = \sum_{n=-\infty}^{\infty} x(nT_s) e^{-j\Omega n T_s}$$

DTFT (Discrete-Time Fourier Transform)

- Now define the discrete-time Fourier transform (DTFT)
- Text calls this "Fourier transform"

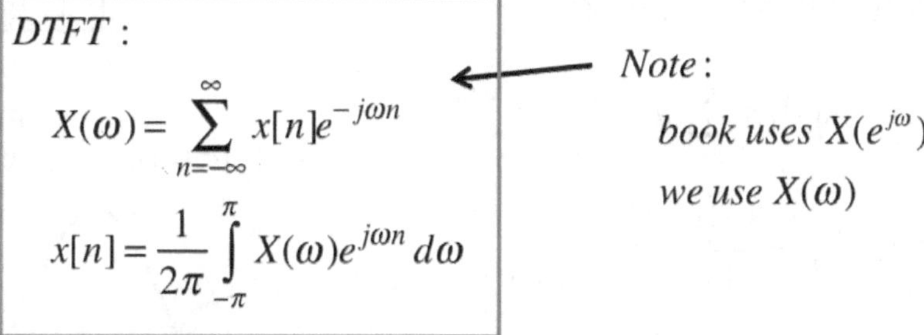

by comparison :

$$X_s(\Omega)=\sum_{n=-\infty}^{\infty} x(nT_s)e^{-j\Omega nT_s}$$

so with $x[n]=x(nT_s)$ *and with* $\omega=\Omega T_s$

$$X(\omega)=X_s(\Omega)\big|_{\Omega=\omega/T_s}$$

Comments on DTFT

- DTFT is continuous in ω
- DTFT is periodic in ω, with period 2π
- Since $\omega=\Omega T_s= 2\pi f/f_s$ then: $\omega= 2\pi$ at $f=f_s$
- Result of DTFT is complex
- Closely related to $X_s(\Omega)$

DTFT :

$$X(\omega)=\sum_{n=-\infty}^{\infty} x[n]e^{-j\omega n}$$

$$x[n]=\frac{1}{2\pi}\int_{-\pi}^{\pi} X(\omega)e^{j\omega n}\,d\omega$$

and $X(\omega)=X_s(\Omega)\big|_{\Omega=\omega/T_s}$

X(Ω) A Ω
X_s(Ω) A/T_s -Ω_s Ω_s Ω
X(ω) A/T_s -2π 2π ω

DTFT Example

- Find DTFT of 3-point moving sum
  y[n]=x[n] + x[n-1] + x[n-2]
  Note: h[n]=u[n]-u[n-3]

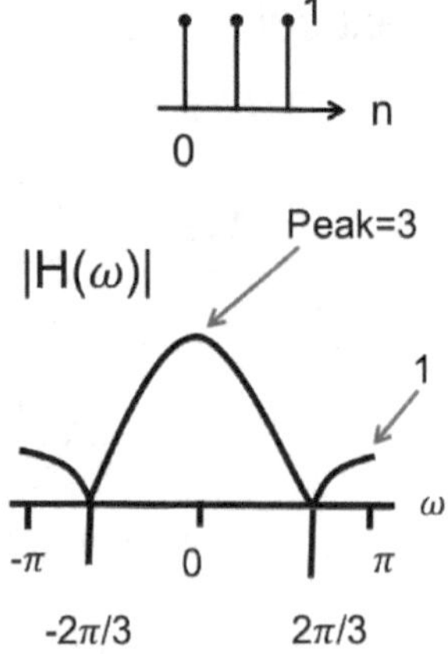

$$H(\omega)=\mathcal{F}\{h[n]\}=\sum_{n=-\infty}^{\infty}h[n]e^{-j\omega n}=\sum_{n=0}^{2}1\cdot e^{-j\omega n}$$

$$=\sum_{n=0}^{2}\left(e^{-j\omega}\right)^{n}=\frac{1-\left(e^{-j\omega}\right)^{3}}{1-e^{-j\omega}}=\frac{1-e^{-j3\omega}}{1-e^{-j\omega}}$$

$$=\frac{e^{-j3\omega/2}\left(e^{j3\omega/2}-e^{-j3\omega/2}\right)}{e^{-j\omega/2}\left(e^{j\omega/2}-e^{-j\omega/2}\right)}=\frac{\sin(3\omega/2)e^{-j\omega}}{\sin(\omega/2)}$$

$$\text{using }\sum_{k=N_1}^{N_2}\alpha^{k}=\frac{\alpha^{N_1}-\alpha^{N_2+1}}{1-\alpha}\quad ;\ N_2>N_1$$

From the difference equation, if the input is x[n]=2, what is the output?

Why does peak=3?
Note: plot is periodic
Period=?
Is this a lowpass filter?
What is dc response?

Note on Sinc() Function Nulls

- The convolution of the rectangle with the two sinusoids at the right would both equal zero, because the area under multiple cycles is zero
- The discrete-time case below would behave the same way

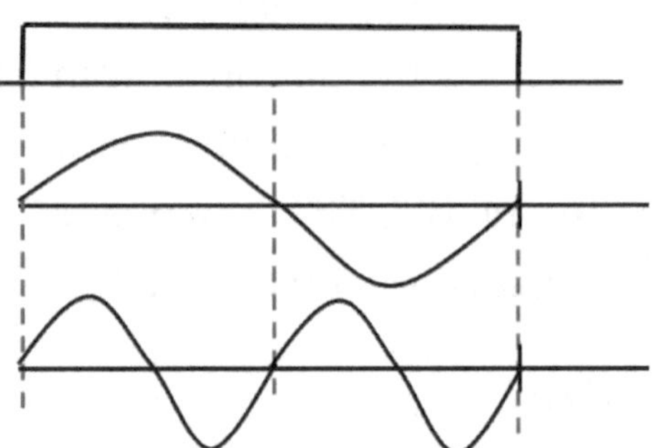

Fourier transform of rectangle would yield a sinc()-like function

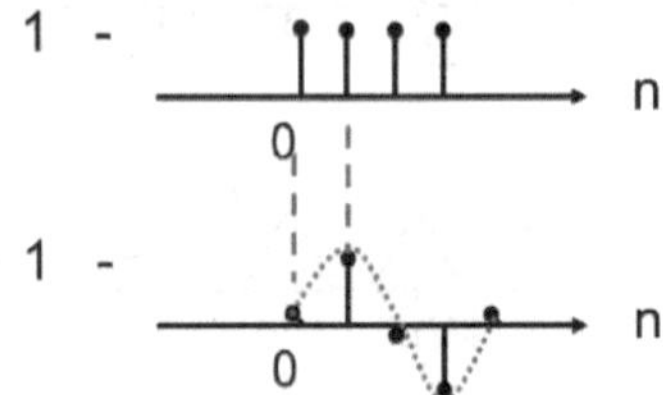

$\mathrm{sinc}(\omega)=\sin(\omega)/\omega$

ω

0

Nulls (zero-crossings)

Eigenfunctions of Convolution

$x[n] = e^{j\omega n}$ → h[n] → y[n]

$$y[n] = h[n] * x[n] = \sum_{k=-\infty}^{\infty} h[k]x[n-k] = \sum_{k=-\infty}^{\infty} h[k]e^{j\omega(n-k)}$$

$$= e^{j\omega n} \sum_{k=-\infty}^{\infty} h[k]e^{-j\omega k} = e^{j\omega n} H(\omega)$$

- Eigenfunction $e^{j\omega n}$
- Eigenvalue $\sum h(\alpha)\, e^{j\omega\alpha}$
- Similarly for sin[ωn], cos[ωn]
- So, DTFT arises naturally from convolution

DTFT Example

- Find DTFT of backward difference

  y[n]=x[n] - x[n-1]

  h[n]= δ[n] - δ[n-1]

1

n

0 -1

$$H(\omega) = \mathcal{F}\{h[n]\} = \sum_{n=-\infty}^{\infty} h[n]e^{-j\omega n}$$

$$= 1 - e^{-j\omega}$$

$$= e^{-j\omega/2}\left(e^{j\omega/2} - e^{-j\omega/2}\right)$$

$$= 2j\sin(\omega/2)e^{-j\omega/2}$$

Using the difference equation, if the input is x[n]=2, what is the output?

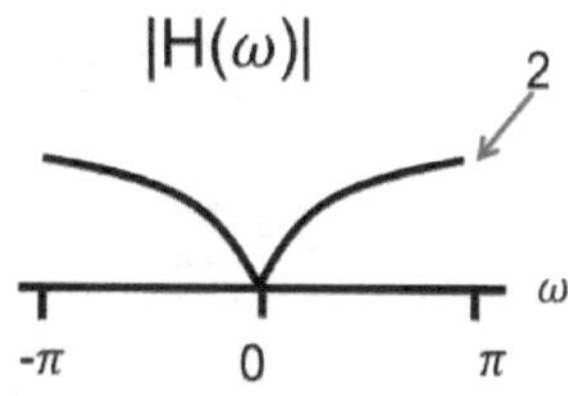

Why does peak=2?
Note: plot is periodic
Period=?
Is this a lowpass filter?
What is dc response?

Properties of DTFT

- Delay/Shift property

$$\mathcal{F}\{x[n-n_0]\} = \sum_{n=-\infty}^{\infty} x[n-n_0]e^{-j\omega n}$$

$$= \sum_{\beta=-\infty}^{\infty} x[\beta]e^{-j\omega(\beta+n_0)} = e^{-j\omega n_0} \sum_{\beta=-\infty}^{\infty} x[\beta]e^{-j\omega\beta}$$

$so:$

$$\mathcal{F}\{x[n-n_0]\} = e^{-j\omega n_0} X(\omega)$$

- Parseval's Theorem

$$Energy = \sum_{n=-\infty}^{\infty} |x[n]|^2 = \frac{1}{2\pi} \int_{-\pi}^{\pi} |X(\omega)|^2 d\omega$$

Properties of DTFT

- Convolution property

$$\mathcal{F}\{a[n]*b[n]\} = \sum_{n=-\infty}^{\infty} \left(a[n]*b[n]\right)e^{-j\omega n}$$

$$= \sum_{n=-\infty}^{\infty} \left(\sum_{k=-\infty}^{\infty} a[k]b[n-k] \right) e^{-j\omega n} = \sum_{k=-\infty}^{\infty} a[k] \left(\sum_{n=-\infty}^{\infty} b[n-k]e^{-j\omega n} \right)$$

$$= \sum_{k=-\infty}^{\infty} a[k]\left(B(\omega)e^{-j\omega k}\right) = B(\omega) \sum_{k=-\infty}^{\infty} a[k]e^{-j\omega k}$$

$so:$

$$\mathcal{F}\{a[n]*b[n]\} = A(\omega)B(\omega)$$

DTFT Across Block Diagram

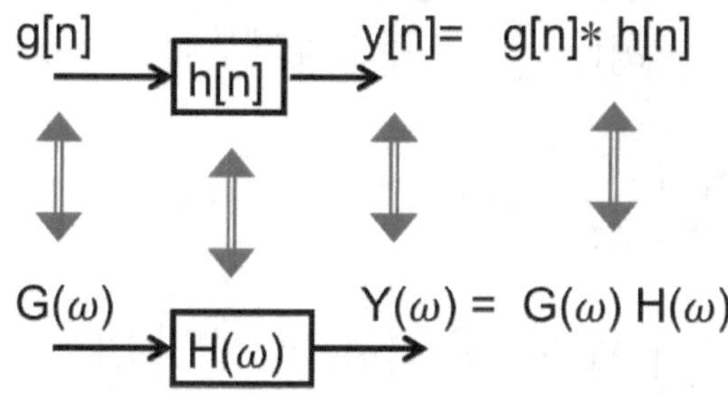

- Take DTFT across block diagram
- Each discrete-time domain element has corresponding frequency domain element
- Two languages have different word for same thing
- Is frequency or time the "true" representation?

Backward Difference Across Block Diagram

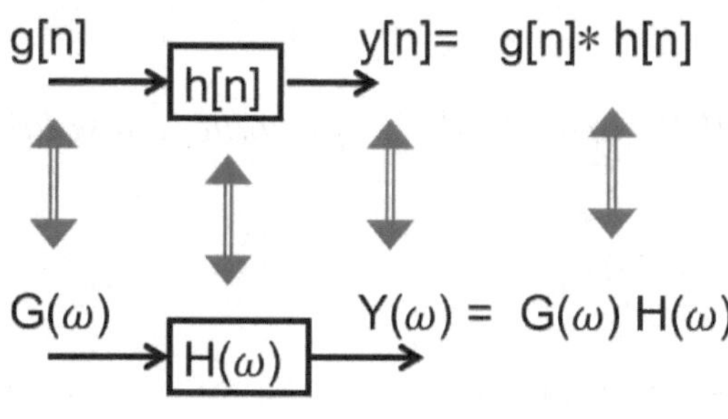

h[n] 1 0 -1 n

$$y[n] = x[n] - x[n-1]$$

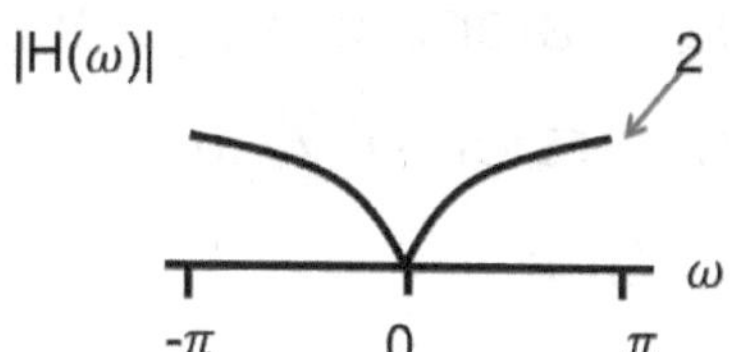

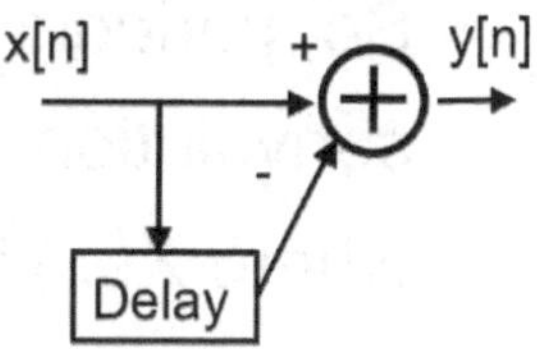

Properties of DTFT

- Modulation (windowing) property

$$\mathcal{F}\{x[n]w[n]\} = \sum_{n=-\infty}^{\infty} x[n]w[n]e^{-j\omega n}$$

$$= \sum_{n=-\infty}^{\infty} \left(\frac{1}{2\pi} \int_{-\pi}^{\pi} X(\beta)e^{j\beta n}\, d\beta \right) w[n]e^{-j\omega n}$$

$$= \frac{1}{2\pi} \int_{-\pi}^{\pi} X(\beta) \left(\sum_{n=-\infty}^{\infty} w[n]e^{-j(\omega-\beta)n} \right) d\beta$$

$$= \frac{1}{2\pi} \int_{-\pi}^{\pi} X(\beta)W(\omega-\beta)\, d\beta \qquad \textit{periodic conv.} \times 1/2\pi$$

Limits not infinite

$so:$

$$\boxed{\mathcal{F}\{x[n]w[n]\} = \frac{1}{2\pi} \int_{-\pi}^{\pi} X(\beta)W(\omega-\beta)\, d\beta}$$

Note on Periodic Convolution

- Periodic convolution from DTFT of multiplication

$$DTFT\{x[n]w[n]\} = \frac{1}{2\pi} \int_{-\pi}^{\pi} X(\beta)W(\omega-\beta)\, d\beta \qquad \textit{periodic convolution}$$

$$= \frac{1}{2\pi} \int_{-\infty}^{\infty} \left\{ \Pi\left(\frac{\beta}{2\pi} \right) X(\beta) \right\} W(\omega-\beta)\, d\beta$$

where $\Pi(a) =$ rectangular box of width 1

- So, periodic convolution is equivalent to "regular convolution" with a truncated version of X(ω), where X(ω) is truncated to $-\pi$ to π

Properties of DTFT

- Frequency shift property

$$\mathcal{F}\left\{m[n]e^{j\omega_c n}\right\} = M(\omega-\omega_c)$$

$$m[n]e^{j\omega_c n} \Leftrightarrow M(\omega-\omega_c)$$

$$\mathcal{F}\left\{m[n]\cos[\omega_c n]\right\} = \sum_{n=-\infty}^{\infty} m[n]\cos[\omega_c n]e^{-j\omega n}$$

$$= \sum_{n=-\infty}^{\infty} \frac{m[n]}{2}\left(e^{j\omega_c n} + e^{-j\omega_c n}\right)e^{-j\omega n} = \sum_{n=-\infty}^{\infty} \frac{m[n]}{2}\left(e^{-j(-\omega_c+\omega)n} + e^{-j(\omega_c+\omega)n}\right)$$

$$= \frac{1}{2}\left[M(\omega-\omega_c) + M(\omega+\omega_c)\right]$$

$$So: \quad m[n]\cos[\omega_c n] \Leftrightarrow \frac{1}{2}\left[M(\omega-\omega_c) + M(\omega+\omega_c)\right]$$

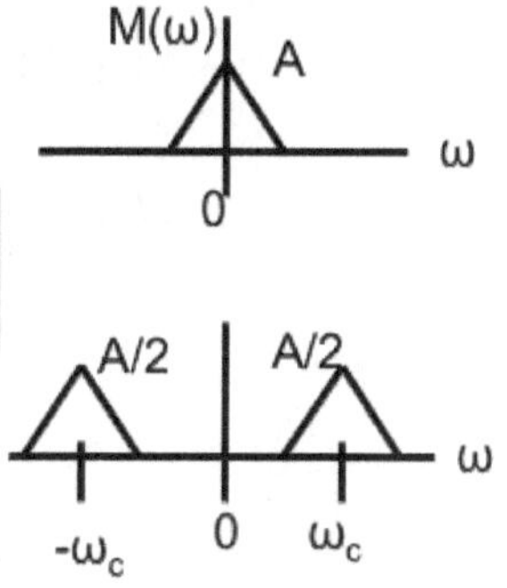

M(ω), A, ω, 0

A/2, A/2, ω, $-\omega_c$, 0, ω_c

- NOTE:
- Frequency shift property of DTFT

$$where\ \mathrm{F}\ \left\{m[n]\right\} = M(\omega)$$

$$\mathrm{F}\ \left\{m[n]e^{j\omega_c n}\right\} = M(\omega-\omega_c)$$

So:

$$m[n]e^{j\omega_c n} \Leftrightarrow M(\omega-\omega_c)$$

and

$$\mathrm{F}\ \left\{m[n]e^{j\pi n}\right\} = \mathrm{F}\ \left\{m[n](-1)^n\right\} = M(\omega-\pi)$$

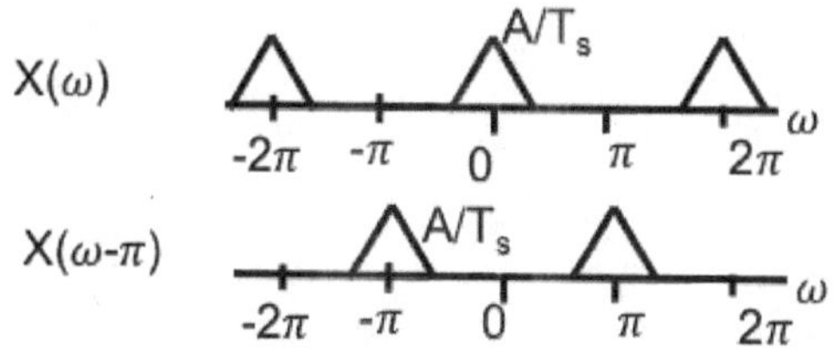

X(ω): A/T_s, ω, -2π, $-\pi$, 0, π, 2π

X(ω−π): A/T_s, ω, -2π, $-\pi$, 0, π, 2π

Using DTFT on Difference Eqns.

- Consider system: y[n] – 0.5 y[n-1] = x[n] – 0.25 x[n-1]
- Take DTFT of both sides of difference equation
- See text example

$$\mathcal{F}\{y[n]-0.5y[n-1]\}=\mathcal{F}\{x[n]-0.25x[n-1]\}$$
$$Y(\omega)-0.5Y(\omega)e^{-j\omega}=X(\omega)-0.25X(\omega)e^{-j\omega}$$
$$Y(\omega)\left(1-0.5e^{-j\omega}\right)=X(\omega)\left(1-0.25e^{-j\omega}\right)$$
$$H(\omega)=\frac{Y(\omega)}{X(\omega)}=\frac{1-0.25e^{-j\omega}}{1-0.5e^{-j\omega}}$$

What is dc response?

5 DISCRETE FOURIER TRANSFORM

The lecture notes in this chapter present the DFT (discrete Fourier transform), wraparound, circular shift, circular convolution, and matrix form of the DFT.

Discrete Fourier Transform

DFT (Discrete Fourier Transform)

- Compare the DTFT to definition of DFT

$DTFT:$

$$X(\omega)=\sum_{n=-\infty}^{\infty} x[n]e^{-j\omega n}$$

$$x[n]=\frac{1}{2\pi}\int_{-\pi}^{\pi} X(\omega)e^{j\omega n}\,d\omega$$

$Note:$

$text\ uses\ X(e^{j\omega})$

$we\ use\ X(\omega)$

Note: $X[k]=X(\omega)|_{\omega=2\pi k/n}$ only if x[n]=0 for n<0 and n>N-1

- Discrete Fourier Transform (DFT)

$DFT:$

$$X[k]=\sum_{n=0}^{N-1} x[n]e^{-j\frac{2\pi}{N}nk}=\sum_{n=0}^{N-1} x[n]W_N^{nk}=X(\omega)\Big|_{\omega=2\pi k/N}$$

$$x[n]=\frac{1}{N}\sum_{k=0}^{N-1} X[k]e^{j\frac{2\pi}{N}nk}$$

Note: Length of sequences x[n] and X[k] are finite (length=N)

Comparison of DFT and DTFT

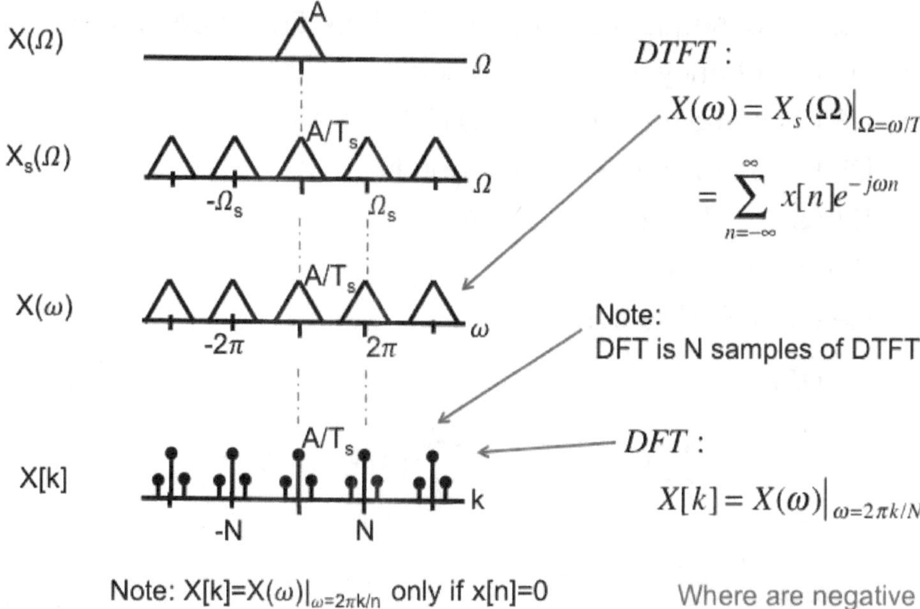

Note: $X[k]=X(\omega)|_{\omega=2\pi k/n}$ only if $x[n]=0$ for $n<0$ and $n>N-1$

Where are negative frequencies ?

DFT & DTFT

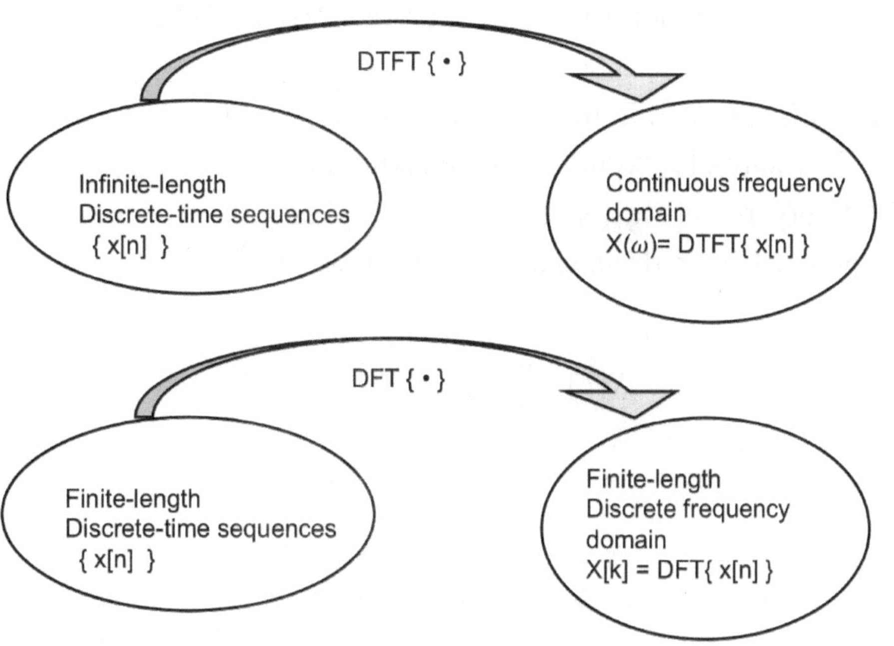

Comments on DFT

- Length of sequences x[n] and X[k] are finite (length=N)
- DFT frequency variable is discrete
- DFT is sampled version of DTFT
- DFT is periodic in k, with period N
- Since $\omega=\Omega T_s= 2\pi f/f_s = 2\pi k/N$
  then: k= N at $\omega=2\pi$ and at $f=f_s$
- Result of DFT is complex

DFT :

$$X[k]=\sum_{n=0}^{N-1}x[n]e^{-j\frac{2\pi}{N}nk}=\sum_{n=0}^{N-1}x[n]W_N^{nk}=X(\omega)\Big|_{\omega=2\pi k/N}$$

$$x[n]=\frac{1}{N}\sum_{k=0}^{N-1}X[k]e^{j\frac{2\pi}{N}nk}$$

FFT (Fast Fourier Transform)

- FFT is fast Fourier transform, Cooley & Tukey 1965
- Result is exactly same as DFT
- Just a faster algorithm for computing DFT
- DFT requires N^2 complex multiplications
- FFT requires $N \log_2(N)$ complex multiplications
- Typically FFT is done with N being power of 2

$$X[k]=\sum_{n=0}^{N-1}x[n]e^{-j\frac{2\pi}{N}nk}$$

For N=1024, $N^2 = 10^6$ and $N \log_2(N) = 10^4$
For N= 10^6, $N^2 = 10^{12}$ and $N \log_2(N) = 2\times10^7$

DFT Example

DFT :

$$X[k]=\sum_{n=0}^{N-1}x[n]e^{-j\frac{2\pi}{N}nk}=X(\omega)\Big|_{\omega=2\pi k/N}$$

$$X[0]=\sum_{n=0}^{4-1}1e^{-j\frac{2\pi}{4}n\cdot 0}=1+1+1+1=4$$

$$X[1]=\sum_{n=0}^{4-1}1e^{-j\frac{2\pi}{4}n\cdot 1}=1-j-1+j=0$$

$$X[2]=\sum_{n=0}^{4-1}1e^{-j\frac{2\pi}{4}n\cdot 2}=1-1+1-1=0$$

$$X[3]=\sum_{n=0}^{4-1}1e^{-j\frac{2\pi}{4}n\cdot 3}=1+j-1-j=0$$

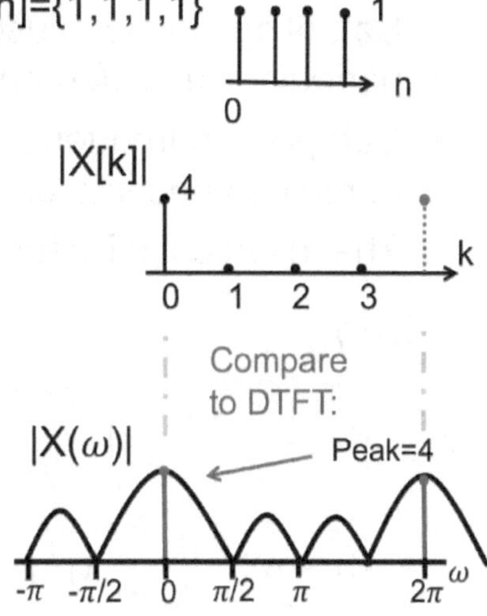

What is dc response?

DFT :

$$X[k]=X(\omega)\Big|_{\omega=2\pi k/N} \quad X(\omega)=\frac{\sin(2\omega)e^{-j3\omega/2}}{\sin(\omega/2)}$$

DFT Example

x[n]={1,0,0,0}

DFT :

$$X[k]=\sum_{n=0}^{N-1}x[n]e^{-j\frac{2\pi}{N}nk}=X(\omega)\Big|_{\omega=2\pi k/N}$$

$$X[0]=e^{-j\frac{2\pi}{4}0\cdot 0}=1$$

$$X[1]=e^{-j\frac{2\pi}{4}0\cdot 1}=1$$

$$X[2]=e^{-j\frac{2\pi}{4}0\cdot 2}=1$$

$$X[3]=e^{-j\frac{2\pi}{4}0\cdot 3}=1$$

1

0 n

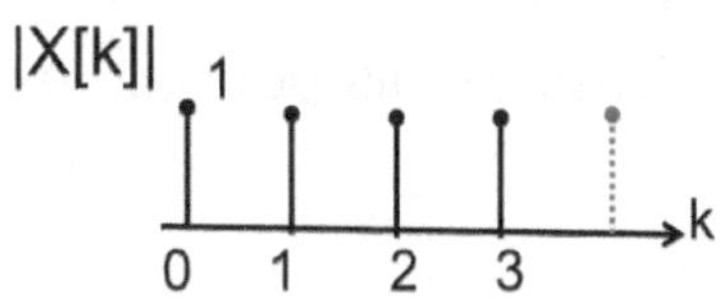

$X(\omega)=1$

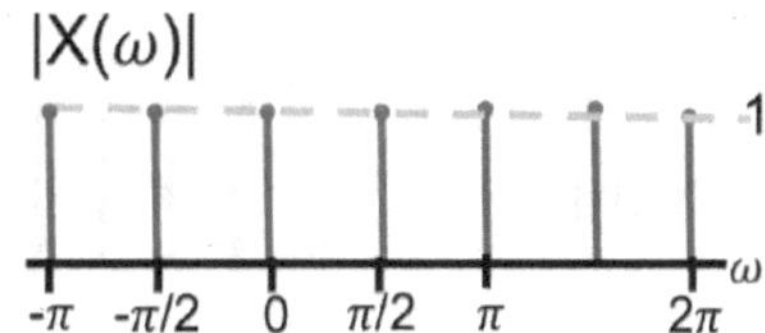

What is dc response?

Periodicity of Inverse DFT

- Sampling in time caused frequency domain to be periodic , i.e., $X_s(\Omega)$ and $X(\omega)$ are periodic
- Sampling in frequency domain causes the discrete-time domain to become periodic
- The inverse DFT causes x[n] to have period N

$$IDFT:$$

$$x[n] = \frac{1}{N}\sum_{n=0}^{N-1} x[n]e^{j\frac{2\pi}{N}nk}$$

$$x[n+N] = \frac{1}{N}\sum_{k=0}^{N-1} X[k]e^{j\frac{2\pi}{N}(n+N)k} = \left(\frac{1}{N}\sum_{k=0}^{N-1} X[k]e^{j\frac{2\pi}{N}nk}\right)e^{j\frac{2\pi}{N}Nk} = x[n]$$

Periodic Extension

- The inverse DFT causes x[n] to be periodic, with period N
- One way to view this is to periodically extend the original N-point sequence x[n]
- Denote this periodic extension $x_p[n]$

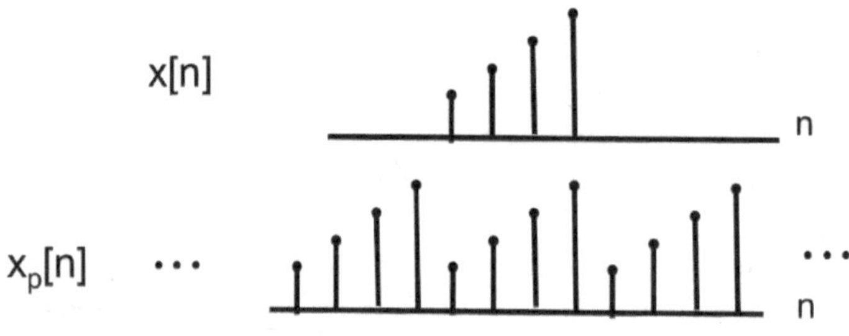

Circular Shift

- A consequence of periodic x[n] is that a shift of the sequence becomes a "circular shift"
- This is illustrated below, where the N-point length signal x[n] is shifted one sample to the right
- The periodic extension results in data being "shifted in from the left"

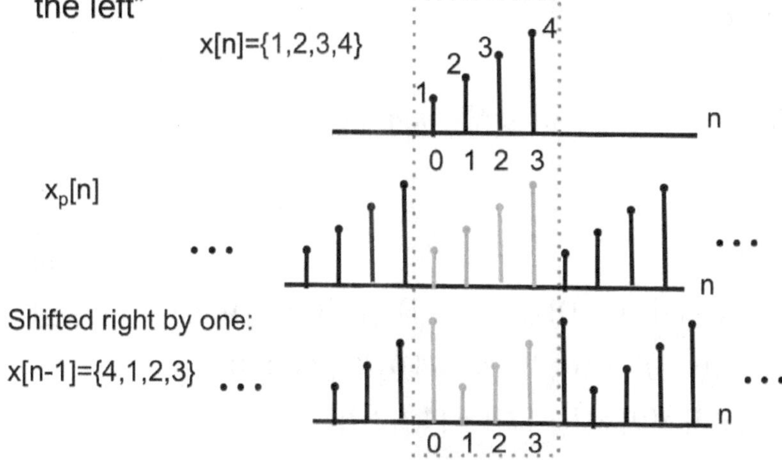

Circular Shift & Wraparound

- Another way to visualize circular shift is wraparound
- If N-point length x[n] is shifted, it "wraps around"

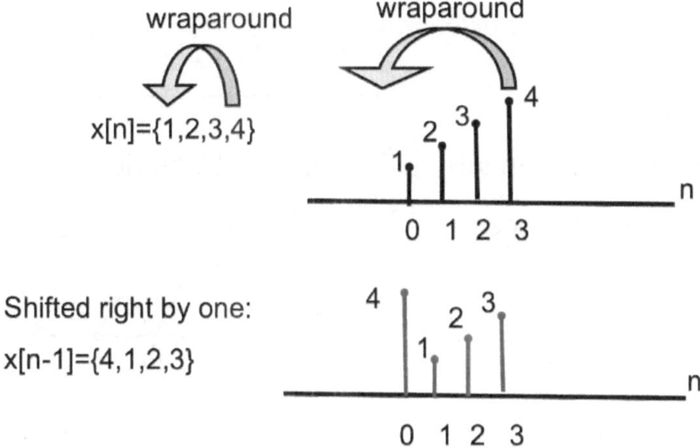

Circular Shift & Modular Arithmetic

- Another way to express circular shift is using modular arithmetic
- "n modulo N" denoted $((n))_N$

Can be thought of as up/down counter.
What if a 2-bit counter is decremented from 00?

Definition:

$$((n))_N = \alpha$$

such that:

$$\alpha + kN = n \quad and \quad 0 \le \alpha < N \quad and \text{ k an integer}$$

count	*bits*
−1	?
0	00
1	01
2	10
3	11
4	00
5	01
6	10

Example

$$x[n] = \{1,2,3,4\}$$

$$x[((n-1))_4] = \{x[((0-1))_4], x[((1-1))_4], x[((2-1))_4], x[((3-1))_4]\}$$

$$= \{x[((-1))_4], x[((0))_4], x[((1))_4], x[((2))_4]\}$$

$$= \{x[3], x[0], x[1], x[2]\} = \{4,1,2,3\}$$

DFT Example, revisited

- Reconsider an earlier DFT example in light of periodic extension of x[n] where:
  - x[n]={1,1,1,1}

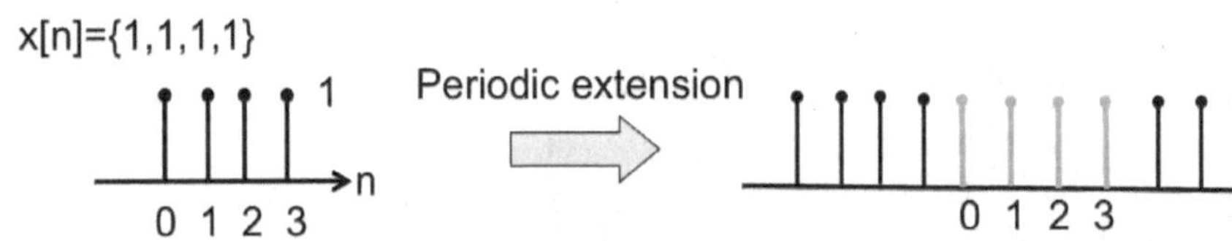

- Periodic extension shows why DFT only contains dc component

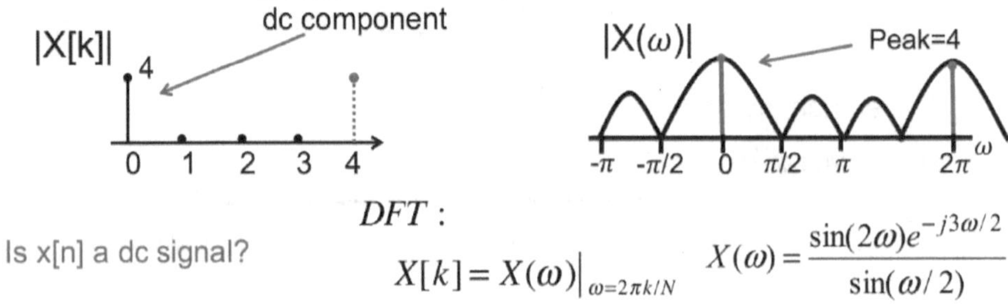

Is x[n] a dc signal?

DFT:

$$X[k] = X(\omega)\big|_{\omega=2\pi k/N} \qquad X(\omega) = \frac{\sin(2\omega)e^{-j3\omega/2}}{\sin(\omega/2)}$$

Circular Convolution

- Recall linear convolution definition:

Linear *convolution* :

$$h[n]*x[n] = \sum_{k=-\infty}^{\infty} h[k]x[n-k]$$

- For DFT, shifts become circular shifts
- Define circular convolution of two sequences of length N:

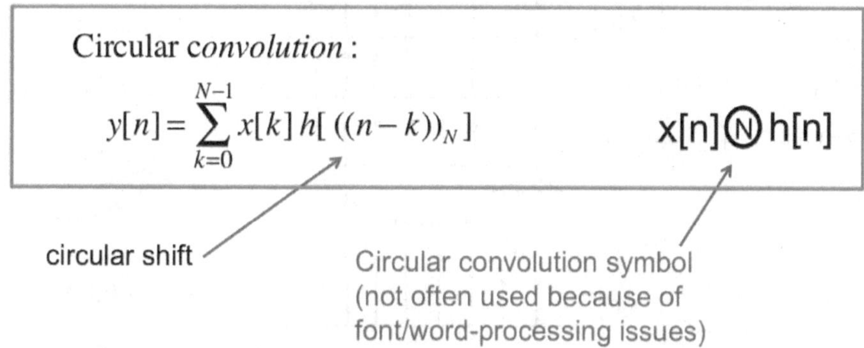

Circular *convolution* :
$y[n] = \sum_{k=0}^{N-1} x[k]\, h[\,((n-k))_N\,]$
x[n]Ⓝh[n]
circular shift
Circular convolution symbol
(not often used because of
font/word-processing issues)

Circular Convolution example (N=8)

- Output y[n] = sum of weighted-delayed impulse responses

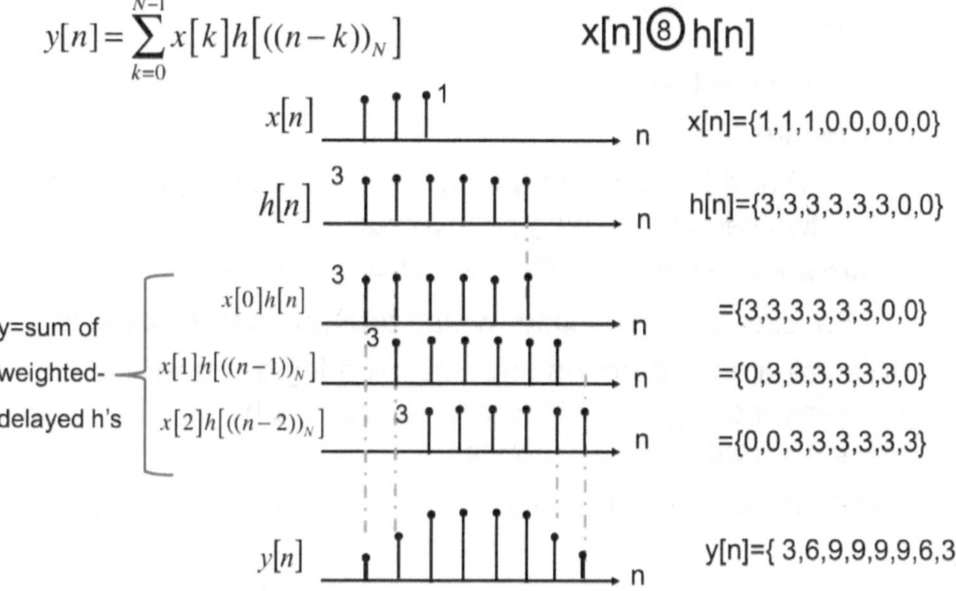

Circular Convolution Example (N=8)

- Output y[n] = sum of weighted-delayed impulse responses

$$y[n]=\sum_{k=0}^{N-1}x[k]h[((n-k))_N]$$

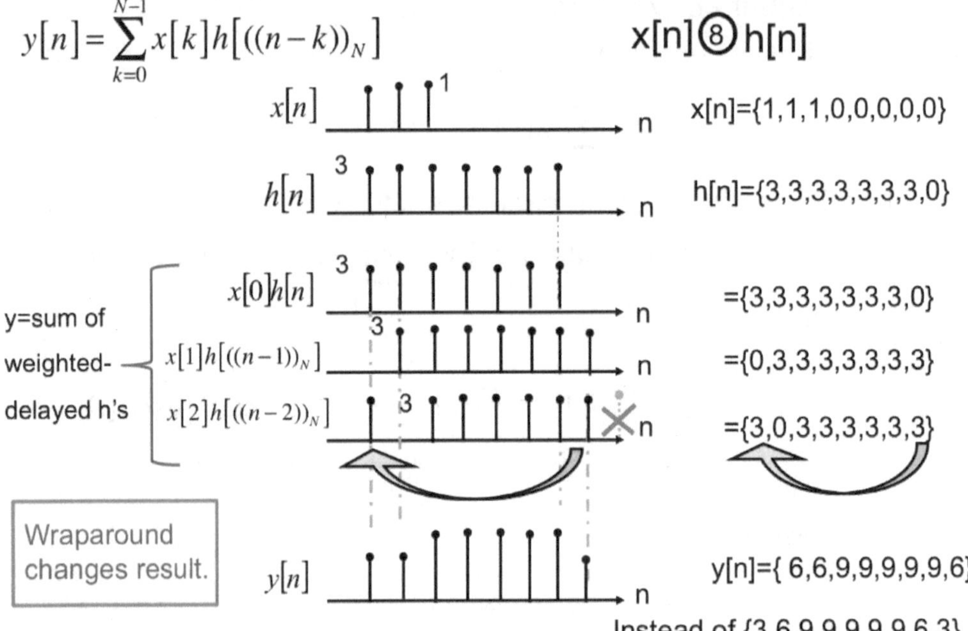

Circular Convolution and Zero-Padding

- Circular convolution gives same result as linear convolution

  If

  N ≥ P + L -1

  where

  P and L are the length of leading nonzero terms in the two sequences being convolved
- In previous example, N=8, P + L -1 = 3+7-1 = 9

  So, circular convolution "wraps around" in this example
- To eliminate wraparound error, zero-padding is used to increase the length N of the two sequences by adding zeroes to end of the sequences
- Use zero-padding to satisfy N ≥ P + L -1

Properties of DFT

- Delay/Shift property (circular shift property)

$$DFT\{x[((n-n_0))_N]\} = \sum_{n=0}^{N-1} x[((n-n_0))_N] e^{-j\frac{2\pi}{N}kn}$$

$$= \sum_{\beta=-n_0}^{N-1-n_0} x[((\beta))_N] e^{-j\frac{2\pi}{N}k(\beta+n_0)} = e^{-j\frac{2\pi}{N}kn_0} \sum_{\beta=-n_0}^{N-1-n_0} x[((\beta))_N] e^{-j\frac{2\pi}{N}k\beta}$$

$$= e^{-j\frac{2\pi}{N}kn_0} \left(\sum_{\beta=0}^{N-1-n_0} x[\beta] e^{-j\frac{2\pi}{N}k\beta} + \sum_{\beta=-n_0}^{-1} x[((\beta))_N] e^{-j\frac{2\pi}{N}k\beta} \right)$$

$$= e^{-j\frac{2\pi}{N}kn_0} \left(\sum_{\beta=0}^{N-1-n_0} x[\beta] e^{-j\frac{2\pi}{N}k\beta} + \sum_{\alpha=N-n_0}^{N-1} x[\alpha] e^{-j\frac{2\pi}{N}k\alpha} \right) = e^{-j\frac{2\pi}{N}kn_0} \left(\sum_{n=0}^{N-1} x[n] e^{-j\frac{2\pi}{N}kn} \right)$$

since x[-a] = x[N-a]

so :

$$\boxed{DFT\{x[((n-n_0))_N]\} = e^{-j\frac{2\pi}{N}kn_0} X[k] = W_N^{kn_0} X[k]} \quad where\ W_N = e^{-j\frac{2\pi}{N}}$$

Properties of DTFT

- Circular convolution property

$$DFT\left\{ \sum_{\alpha=0}^{N-1} x[\alpha] h[((n-\alpha))_N] \right\} = \sum_{n=0}^{N-1} \left(\sum_{\alpha=0}^{N-1} x[\alpha] h[((n-\alpha))_N] \right) e^{-j\frac{2\pi}{N}kn}$$

$$= \sum_{\alpha=0}^{N-1} x[\alpha] \left(\sum_{n=0}^{N-1} h[((n-\alpha))_N] e^{-j\frac{2\pi}{N}kn} \right)$$

$$= \sum_{\alpha=0}^{N-1} x[\alpha] \left(e^{-j\frac{2\pi}{N}k\alpha} H[k] \right) = H[k] \sum_{\alpha=0}^{N-1} x[\alpha] e^{-j\frac{2\pi}{N}k\alpha}$$

$$= X[k]H[k]$$

$$\boxed{x[n] \circledN h[n] \quad \Leftrightarrow \quad X[k]H[k]}$$

Convolution Using FFT

- Consider that convolution requires N^2 complex multiplies

$$y[n] = \sum_{k=0}^{N-1} x[k]h[((n-k))_N] \qquad x[n] \circledN h[n]$$

- FFT requires N $\log_2$(N) complex multiplies
- Convolution can be performed in frequency domain as:
  - IFFT{ FFT{x[n]} FFT{y[n]} }
- So, convolution with FFT requires 3(N $\log_2$(N)) + N complex multiplications
- For example:
  - For N=1024, $N^2 = 10^6$ and 3N $\log_2$(N) +N= 3×10^4
  - For N= 10^6, $N^2 = 10^{12}$ and 3N $\log_2$(N) + N = 6×10^7
- So, FFT can implement convolution approximately 10,000 times faster when N=10^6 points

Matrix Form of DFT

- The DFT can also be expressed matrix form (not in text)

$DFT:$

$$X[k]=\sum_{n=0}^{N-1}x[n]e^{-j\frac{2\pi}{N}nk}=\sum_{n=0}^{N-1}x[n]W_N^{nk}$$

$$\begin{bmatrix}X[0]\\X[1]\\X[2]\\X[3]\end{bmatrix}=\begin{bmatrix}1&1&1&1\\1&W_4^1&W_4^2&W_4^3\\1&W_4^2&W_4^4&W_4^6\\1&W_4^3&W_4^6&W_4^9\end{bmatrix}\begin{bmatrix}x[0]\\x[1]\\x[2]\\x[3]\end{bmatrix}=\begin{bmatrix}1&1&1&1\\1&e^{-j\pi/2}&e^{-j\pi}&e^{-j3\pi/2}\\1&e^{-j\pi}&e^{-j2\pi}&e^{-j3\pi}\\1&e^{-j3\pi/2}&e^{-j6\pi/2}&e^{-j9\pi/2}\end{bmatrix}\begin{bmatrix}x[0]\\x[1]\\x[2]\\x[3]\end{bmatrix}$$

$$\begin{bmatrix}X[0]\\X[1]\\X[2]\\X[3]\end{bmatrix}=\begin{bmatrix}1&1&1&1\\1&-j&-1&j\\1&-1&1&-1\\1&j&-1&-j\end{bmatrix}\begin{bmatrix}x[0]\\x[1]\\x[2]\\x[3]\end{bmatrix}$$

$so:$

$$\overline{X}=\overline{\overline{W}}\,\overline{x}$$

How many multiplications are in the above?

Matrix Form of DFT

- The matrix form gives a powerful generalization and insight into more general types of transforms
- Here, the DFT and inverse transform are viewed as matrix multiplications and matrix inverses

$$DFT:\quad \overline{X}=\overline{\overline{W}}\,\overline{x}$$

$$inverse:\quad \overline{x}=\overline{\overline{W}}^{-1}\overline{X}=\frac{1}{N}\overline{\overline{W}}^{*}\,\overline{X}$$

$$because:\quad \overline{x}=\overline{\overline{W}}^{-1}\overline{X}=\overline{\overline{W}}^{-1}\overline{\overline{W}}\,\overline{x}$$

- In a general sense, a transform is a matrix with an inverse
- Advanced signal processing courses may explore this further
- Consider Hadamard transform:

$$\overline{X}=\overline{\overline{H}}\,\overline{x}=\frac{1}{2}\begin{bmatrix}1&1&1&1\\1&-1&1&-1\\1&1&-1&-1\\1&-1&-1&1\end{bmatrix}\overline{x}\quad and\quad \overline{x}=\overline{\overline{H}}^{-1}\overline{X}=\frac{1}{2}\begin{bmatrix}1&1&1&1\\1&-1&1&-1\\1&1&-1&-1\\1&-1&-1&1\end{bmatrix}\overline{X}$$

Matrix DFT Example

- Let x[n]={1,1,1,1}

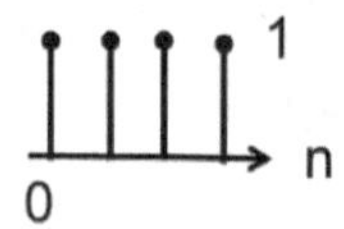

$DFT:$

$$\begin{bmatrix} X[0] \\ X[1] \\ X[2] \\ X[3] \end{bmatrix} = \begin{bmatrix} 1 & 1 & 1 & 1 \\ 1 & -j & -1 & j \\ 1 & -1 & 1 & -1 \\ 1 & j & -1 & -j \end{bmatrix} \begin{bmatrix} x[0] \\ x[1] \\ x[2] \\ x[3] \end{bmatrix}$$

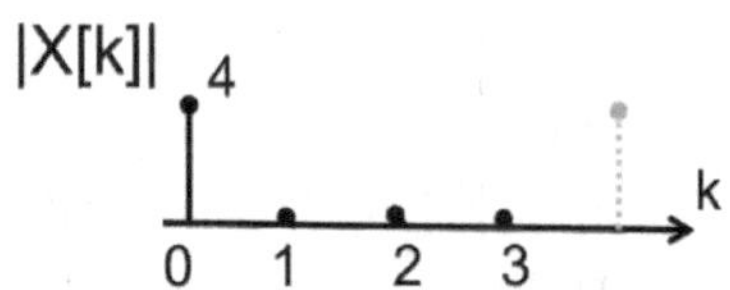

$$= \begin{bmatrix} 1 & 1 & 1 & 1 \\ 1 & -j & -1 & j \\ 1 & -1 & 1 & -1 \\ 1 & j & -1 & -j \end{bmatrix} \begin{bmatrix} 1 \\ 1 \\ 1 \\ 1 \end{bmatrix} = \begin{bmatrix} 4 \\ 0 \\ 0 \\ 0 \end{bmatrix}$$

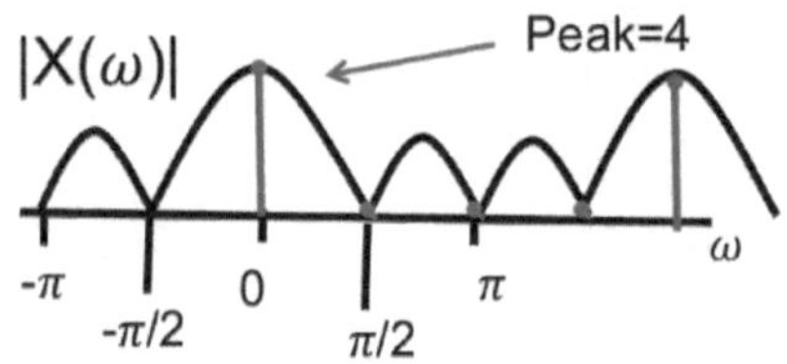

$$X(\omega) = \frac{\sin(2\omega)e^{-j3\omega/2}}{\sin(\omega/2)}$$

6 Z-TRANSFORM

The lecture notes in this chapter cover the z-transform, ROC (region of convergence), and properties of the z-transform.

Z-Transform

Z- Transform – Simplified

- To compute the z-transform of a signal g[n], simply multiply g[n] by powers of z, as illustrated below

Definition:

$$G(z) = Z\{g[n]\} = \sum_{n=-\infty}^{\infty} g[n] z^{-n}$$

inverse:

$$g[n] = \frac{1}{2\pi j} \oint G(z) z^{n-1}\, dz$$

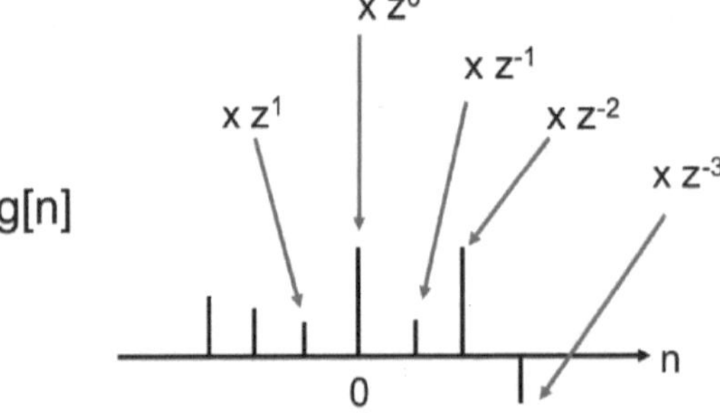

Simply multiply data by corresponding powers of z

Z- Transform Relation to DTFT

- The z-transform is related to DTFT

$$X(z) = Z\{x[n]\} = \sum_{n=-\infty}^{\infty} x[n] z^{-n}$$

$$X(\omega) = DTFT\{x[n]\} = \sum_{n=-\infty}^{\infty} x[n] e^{-j\omega n} = \sum_{n=-\infty}^{\infty} x[n] (e^{j\omega})^{-n}$$

$so:$

$$\boxed{X(\omega) = X(z)\big|_{z=e^{j\omega}}}$$

- So, DTFT corresponds to points on unit circle in the z-plane

Z- Transform Visualization

Note:

X(z) is a function of z that itself could be complex-valued.

Consider to plot |X(z)| we would need a "3-D plot" where the height above the z-plane would correspond to |X(z)|.

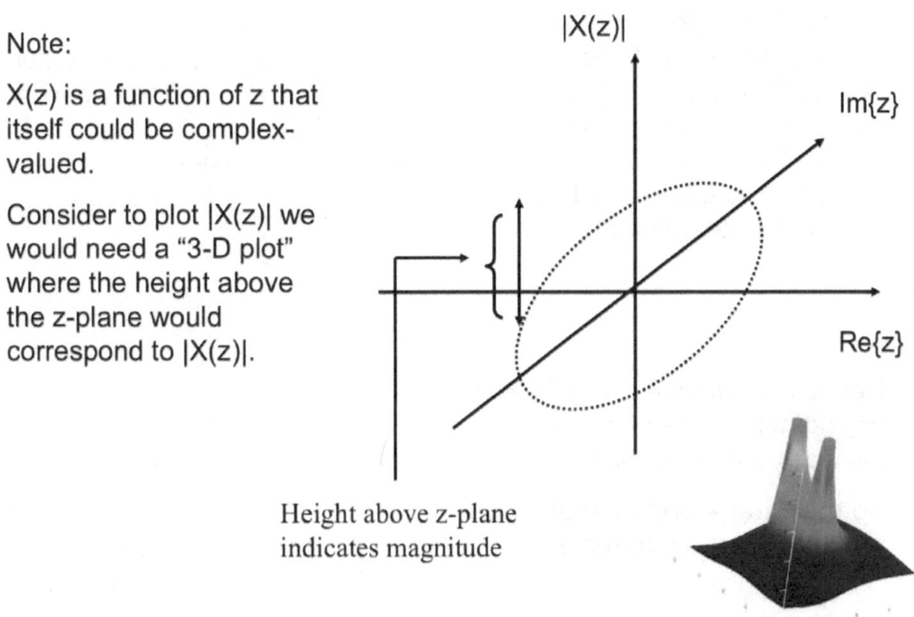

Z- Transform Relation to Laplace Transform

- Z-transform is related to Laplace transform (both 2-sided here)

$$X(z) = Z\{x[n]\} = \sum_{n=-\infty}^{\infty} x[n] z^{-n}$$

x[n]

n

0

$$Define: \quad x_s(t) = \sum_{n=-\infty}^{\infty} x[n]\,\delta(t - nT_s)$$

$x_S(t)$

t

0 T_S

$$X_s(s) = L\{x_s(t)\} = \int_{-\infty}^{\infty} \left(\sum_{n=-\infty}^{\infty} x[n]\,\delta(t - nT_s) \right) e^{-st}\, dt$$

Both 2-sided

$$= \sum_{n=-\infty}^{\infty} x[n] \int_{-\infty}^{\infty} \delta(t - nT_s) e^{-st}\, dt = \sum_{n=-\infty}^{\infty} x[n] e^{-snT_s}$$

$so:$

$$\boxed{X_s(s) = X(z)\big|_{z=e^{sT_s}}}$$

So, $X_S(j\Omega T_S)$ corresponds to points on unit circle in z-plane where $z = e^{j\omega} = e^{j\Omega T_0}$

Im{z} j

$e^{j\omega}$

ω

-1

Re{z}

+1

- j

Z- Transform and Laplace Transform

So: the Z-transform is seen to be "more general" than the DTFT since the DTFT corresponds to the points on the unit circle in the complex plane (i.e: DTFT is a subset of z-transform)

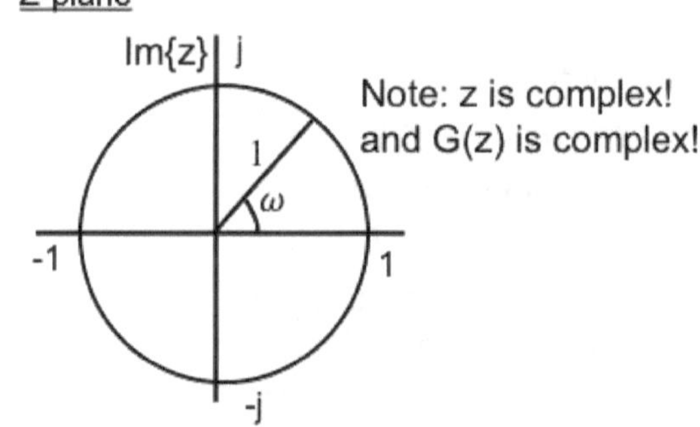

Recall: in continuous-time Fourier transform is on the y-axis of Laplace transform s-plane

<u>So the z-transform is analogous to the Laplace Transform</u>

$Laplace:$

$$x(s) = \int_0^{\infty} e^{-st} x(t)\, dt$$

Laplace s-plane, $s=\sigma+j\Omega$

Im{s}= Ω

Re{s}=σ

Z- Transform

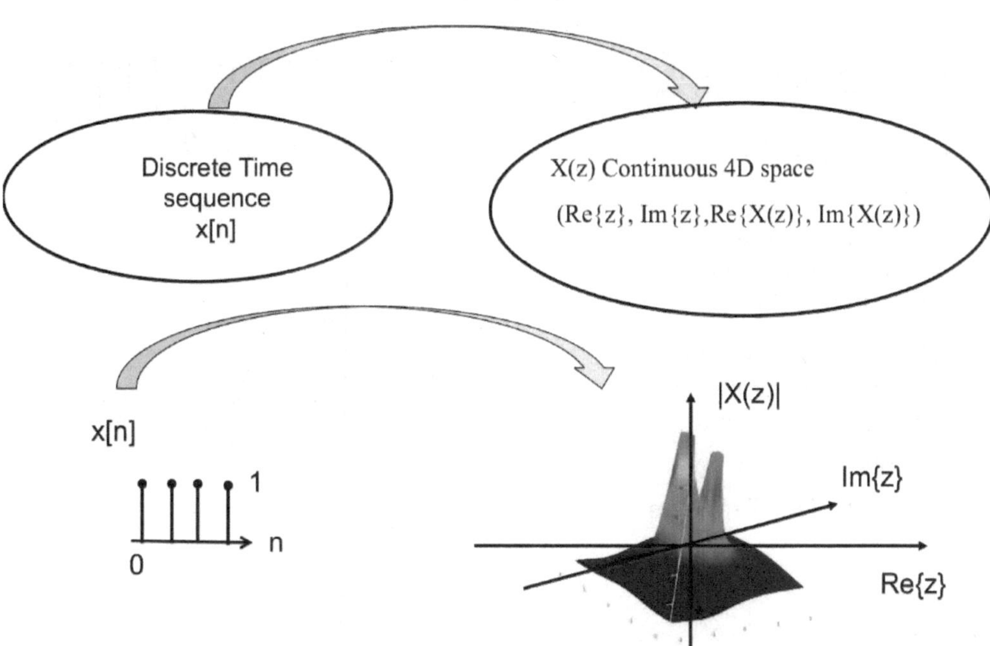

Example Z-Transform

x[n]={1,1,1,1}

- Z-transform

$$X(z) = \sum_{n=-\infty}^{\infty} x[n]z^{-n} = \sum_{n=0}^{3} 1z^{-n} = \frac{1-z^{-4}}{1-z^{-1}}$$

- Now consider DTFT

$$X(\omega) = \sum_{n=-\infty}^{\infty} x[n]e^{-j\omega n} = \sum_{n=0}^{3} 1\cdot e^{-j\omega n} = \frac{1-\left(e^{-j\omega}\right)^4}{1-e^{-j\omega}}$$

- And _this equals the result derived from z-transform above_:

$$X(\omega) = X(z)\big|_{z=e^{j\omega}} = \left.\frac{1-z^{-4}}{1-z^{-1}}\right|_{z=e^{j\omega}} = \frac{1-(e^{j\omega})^{-4}}{1-(e^{j\omega})^{-1}} = \frac{1-e^{-j4\omega}}{1-e^{-j\omega}}$$

And the magnitude is same as before:

$$|X(\omega)| = \left|\frac{1-\left(e^{-j\omega}\right)^4}{1-e^{-j\omega}}\right| = \left|\frac{e^{-j2\omega}\left(e^{j2\omega}-e^{-j2\omega}\right)}{e^{-j\omega/2}\left(e^{j\omega/2}-e^{-j\omega/2}\right)}\right| = \left|\frac{\sin(2\omega)e^{-j3\omega/2}}{\sin(\omega/2)}\right| = \left|\frac{\sin(2\omega)}{\sin(\omega/2)}\right|$$

Note on previous example

$$X(z)=\sum_{n=-\infty}^{\infty}x[n]z^{-n}=\sum_{n=0}^{3}1z^{-n}=\frac{1-z^{-4}}{1-z^{-1}}$$

x[n]={1,1,1,1}

1

n

0

$$and \quad X(\omega)=X(z)\big|_{z=e^{j\omega}}$$

Z-Transform was evaluated on the unit circle to find DTFT

Note: result is periodic on unit circle z= e $^{j\omega}$ with period in ω of 2π as needed for DTFT

Every 2π, DTFT repeats

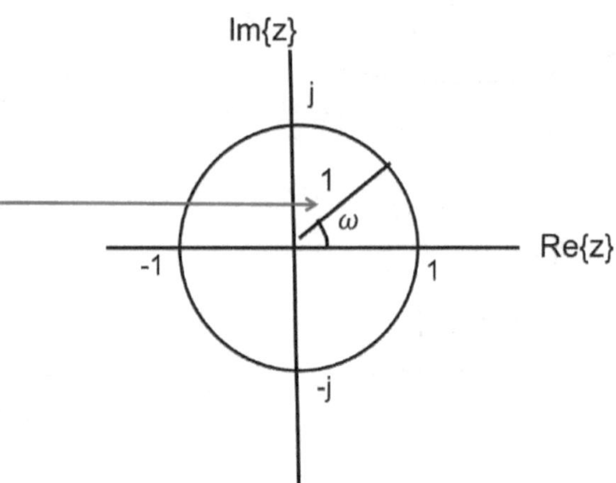

Example Z-Transform

- Find the z-transform of the unit step: x[n] = u[n]
- Z-transform

$$X(z)=\sum_{n=-\infty}^{\infty}x[n]z^{-n}=\sum_{n=0}^{\infty}u[n]z^{-n}=\frac{1-z^{-\infty}}{1-z^{-1}}$$

$$X(z)=\frac{1}{1-z^{-1}}; \quad ROC \ |z|>1$$

- ROC is the region of convergence
  - It is the region in z-plane where X(z) is bounded
  - X(z) is undefined outside the ROC
- So, above ROC is |z| > 1

Note: Why does ROC exclude z=1?

$$\text{hint}: X(z)\big|_{z=1}=\sum_{n=-\infty}^{\infty}x[n]z^{-n}=\sum_{n=0}^{\infty}1=\infty$$

ROC (Region of Convergence)

- ROC is the region in z-plane where the z-transform of a sequence exists (is finite)
- The z-transform X(z) exists if:

$$|X(z)| \leq \left| \sum_{n=-\infty}^{\infty} x[n]z^{-n} \right| \leq \sum_{n=-\infty}^{\infty} \left| x[n]z^{-n} \right| < \infty$$

- In polar notation, $z = r\,e^{j\omega}$, and the z-transform exists if:

$$|X(z)| = \left| \sum_{n=-\infty}^{\infty} x[n]r^{-n}e^{-j\omega n} \right|$$

$$\leq \sum_{n=-\infty}^{\infty} \left| x[n]r^{-n}e^{-j\omega n} \right| \leq \sum_{n=-\infty}^{\infty} \left| x[n]r^{-n} \right| < \infty$$

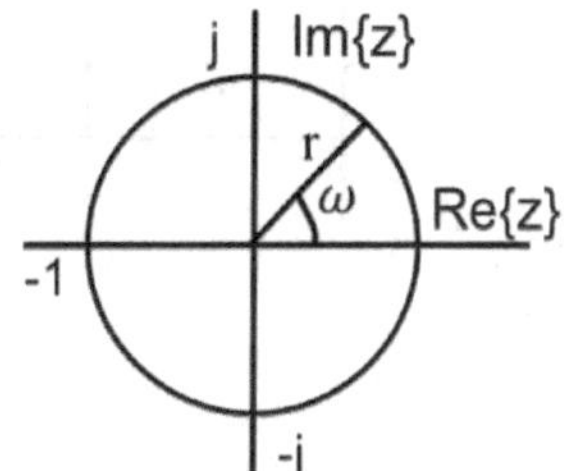

- So the flexibility to select r in the z-transform allows convergence for some sequences for which the DTFT may not exist (r=1 for DTFT).

Region of convergence

- From the previous slide, X(z) exists if:

$$\sum_{n=-\infty}^{\infty} \left| x[n]z^{-n} \right| = \sum_{n=-\infty}^{\infty} \left| x[n]r^{-n} \right| < \infty$$

- So X(z) exists if:

$$\sum_{n=-\infty}^{-1} \left| x[n]r^{-n} \right| + \sum_{n=0}^{\infty} \left| x[n]r^{-n} \right| < \infty$$

- Or X(z) exists if:

$$\underbrace{\sum_{n=1}^{\infty} \left| x[-n]r^{n} \right|}_{\text{Left-side}} + \underbrace{\sum_{n=0}^{\infty} \left| x[n]r^{-n} \right|}_{\text{Right-side}} < \infty$$

"Left-side" or "Non-causal" portion of x[n] This part converges for $r<r_-$

"Right-side" or "Causal" portion of x[n]. This part converges for $r>r_+$

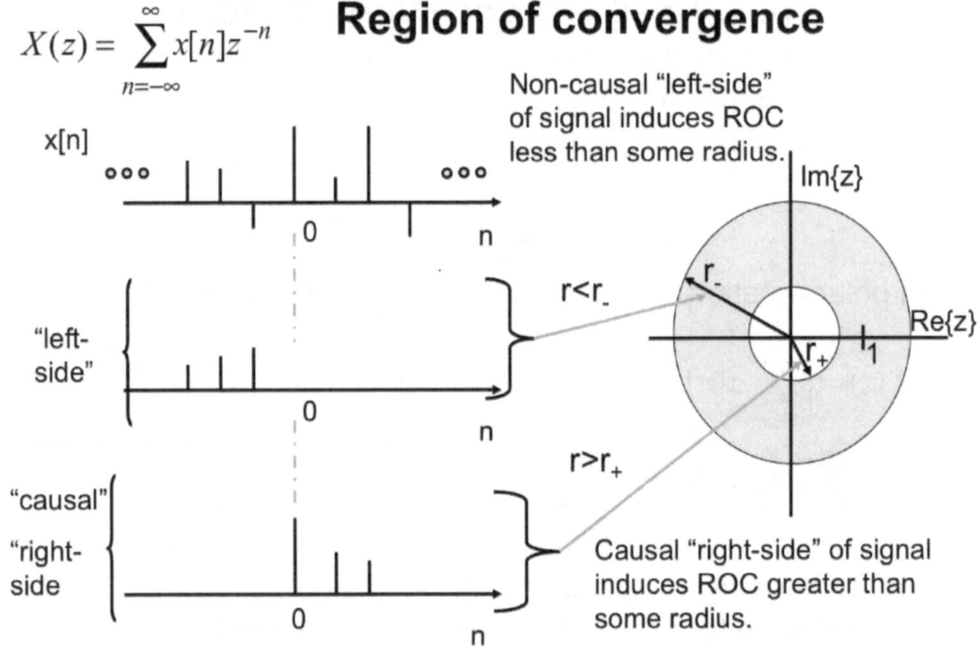

Total ROC is intersection of both ROC, yellow ring above

Example: Exponential Sequence

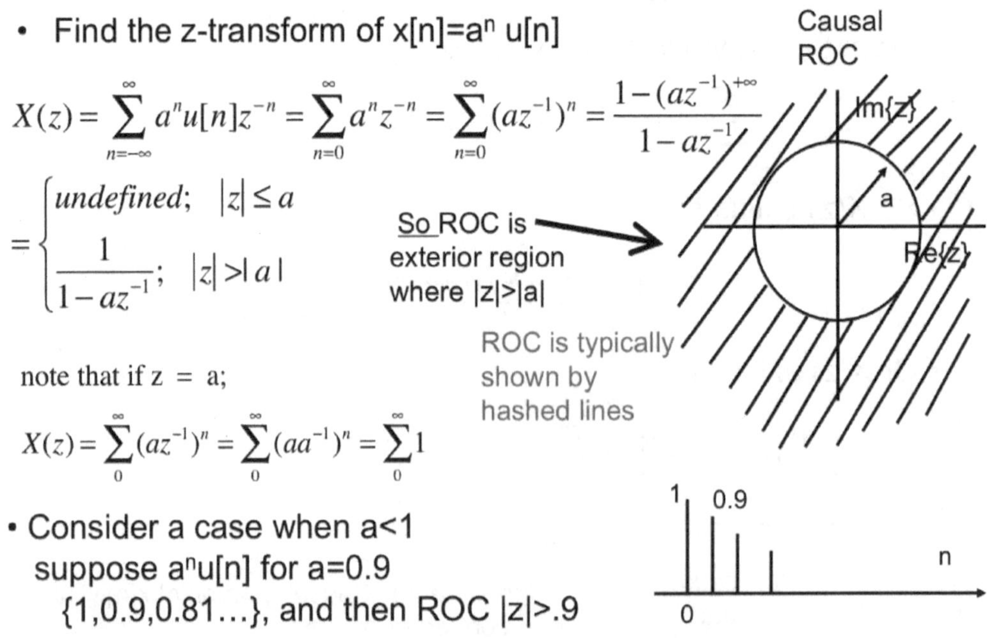

• Find the z-transform of x[n]=a^n u[n]
$X(z) = \sum_{n=-\infty}^{\infty} a^n u[n] z^{-n} = \sum_{n=0}^{\infty} a^n z^{-n} = \sum_{n=0}^{\infty} (az^{-1})^n = \frac{1-(az^{-1})^{+\infty}}{1-az^{-1}}$
$= \begin{cases} undefined; & |z| \le a \\ \frac{1}{1-az^{-1}}; & |z| > |a| \end{cases}$
Causal ROC
So ROC is exterior region where |z|>|a|
Im{z}
a
Re{z}
ROC is typically shown by hashed lines
note that if z = a;
$X(z) = \sum_{0}^{\infty} (az^{-1})^n = \sum_{0}^{\infty} (aa^{-1})^n = \sum_{0}^{\infty} 1$
• Consider a case when a<1 suppose a^n u[n] for a=0.9 {1,0.9,0.81…}, and then ROC |z|>.9
1
0.9
n
0

In this example (a=0.9) unit circle lies in ROC. (DTFT exists).

Two-Sided Z-Transform Example

- Find the z-transform of x[n]=a^n u[n] - b^n u[-n-1]

Left-sided part of signal

$$X(z)=\sum_{n=-\infty}^{\infty} x[n]z^{-n}=-\sum_{n=-\infty}^{-1} b^n z^{-n}+\sum_{n=0}^{\infty} a^n z^{-n}$$

$$=-\sum_{n=0}^{\infty} b^{-1-n} z^{1+n}+\sum_{n=0}^{\infty} a^n z^{-n}$$

$$=-b^{-1}z\sum_{n=0}^{\infty}\left(b^{-1}z\right)^n+\sum_{n=0}^{\infty}(az^{-1})^n$$

$$=-b^{-1}z\left(\frac{1-(b^{-1}z)^{\infty}}{1-b^{-1}z}\right)+\frac{1-(az^{-1})^{\infty}}{1-az^{-1}}$$

$$=-b^{-1}z\left(\frac{1}{1-b^{-1}z}\right)+\frac{1}{1-az^{-1}}$$

$$=\frac{1}{1-bz^{-1}}+\frac{1}{1-az^{-1}}$$

$$ROC\,|z|<|b|\,and\,|z|>|a|$$

Im{z}
a
b
Re{z}

Example if a and b are positive real.

ROC is a ring

Note: same forms, except for ROC

Z-Transform MUST Include ROC

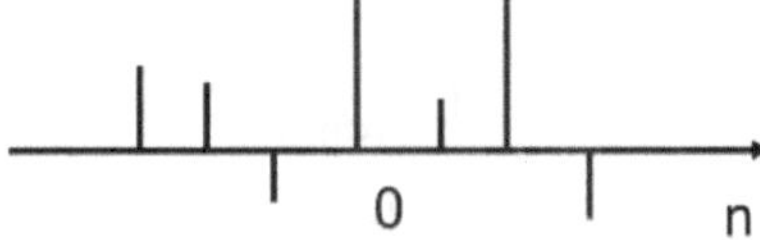

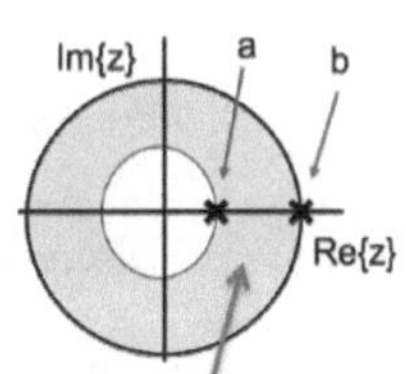

- For the 2-sided example, the left-sided and right-side components of the z-transform had the same form
- See z-transform table in text

$$a^n u[n] \Leftrightarrow \frac{1}{1-az^{-1}};\quad |z|>|a|$$

$$-a^n u[-n-1] \Leftrightarrow \frac{1}{1-az^{-1}};\quad |z|<|a|$$

Note that poles are found at boundary of ROC

- The only difference in the two forms above are the ROC!!
- So, **z-transform must include BOTH X(z) and the ROC**!!

Zeroes of Two-Sided Example

- To find the zeroes of the z-transform of the previous example, rearrange the result:

$$X(z)=\frac{1}{1-bz^{-1}}+\frac{1}{1-az^{-1}}$$
$$=\frac{z}{z-b}+\frac{z}{z-a}$$
$$=\frac{z(z-a)+z(z-b)}{(z-a)(z-b)}$$
$$=\frac{z(2z-a-b)}{(z-a)(z-b)}$$

2 zeroes: z=0 & z=(a+b)/2

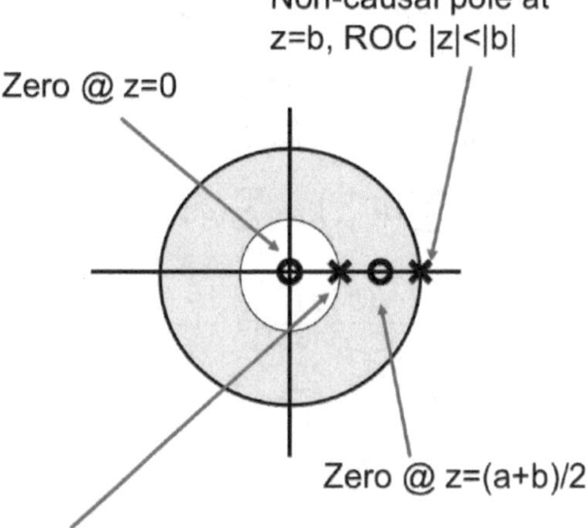

Table of Z-Transforms

Z-transform Pairs

Discrete-time Function	z-transform
$\delta[n]$	1
$u[n]$	$\frac{z}{z-1};\ \lvert z\rvert>1$
$nu[n]$	$\frac{z}{(z-1)^2};\ \lvert z\rvert>1$
$a^n u[n]$	$\frac{z}{z-a};\ \lvert z\rvert>\lvert a\rvert$
$-a^n u[-n-1]$	$\frac{z}{z-a};\ \lvert z\rvert<\lvert a\rvert$
$na^n u[n]$	$\frac{az}{(z-a)^2};\ \lvert z\rvert>\lvert a\rvert$
$\cos(\omega_0 n)u[n]$	$\frac{z^2-z\cos(\omega_0)}{z^2-2z\cos(\omega_0)+1};\ \lvert z\rvert>1$
$\sin(\omega_0 n)u[n]$	$\frac{z\sin(\omega_0)}{z^2-2z\cos(\omega_0)+1};\ \lvert z\rvert>1$

Properties of Z-Transform

1. Linearity:

$$ax_1[n]+bx_2[n] \Leftrightarrow aX_1(z)+bX_2(z)$$

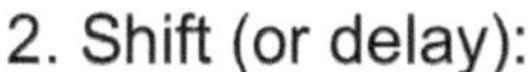

$$Z\{x[n-n_o]\} = \sum_{n=-\infty}^{\infty} x[n-n_o]z^{-n}$$

$$= \sum_{\alpha=-\infty}^{\infty} x[\alpha]z^{-(\alpha+n_o)}$$

$$= z^{-n_o}\sum_{\alpha=-\infty}^{\infty} x[\alpha]z^{-\alpha} = z^{-n_o}X(z)$$

so:

$$x[n-n_o] \quad \Leftrightarrow \quad z^{-n_o}X(z)$$

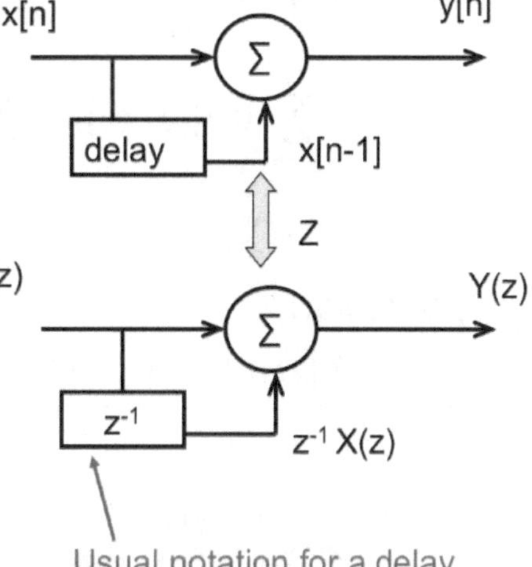

Properties of z-transform

3. Convolution:

$$Z\{x[n]*h[n]\} = \sum_{n=-\infty}^{\infty}\left(\sum_{k=-\infty}^{\infty} x[k]h[n-k]\right)z^{-n}$$

$$= \sum_{k=-\infty}^{\infty} x[k]\left(\sum_{n=-\infty}^{\infty} h[n-k]z^{-n}\right)$$

$$= \sum_{k=-\infty}^{\infty} x[k]z^{-k}H(z) = H(z)\sum_{k=-\infty}^{\infty} x[k]z^{-k}$$

$$= H(z)X(z)$$

so:

$$x[n]*h[n] \Leftrightarrow X(z)H(z)$$

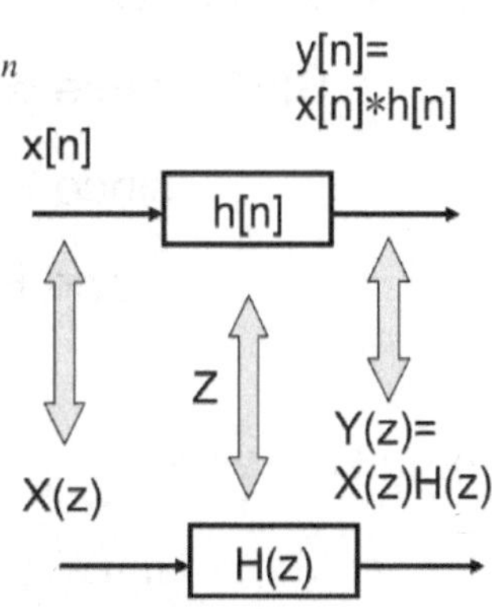

Can take z-transform across block diagram representing any point in either domain

- ROC is usually intersection of ROCs

Properties of z-transform

4. Exponential modulation property:

$$z_o^n x[n] \Leftrightarrow X(z/z_o)$$

$$Z\{z_o^n x[n]\} = \sum_{n=-\infty}^{\infty} z_o^n x[n] z^{-n}$$

$$= \sum_{n=-\infty}^{\infty} x[n]\left(z^{-1}z_o\right)^n = \sum_{n=-\infty}^{\infty} x[n]\left(\frac{z}{z_o}\right)^{-n}$$

$$= X(z)\Big|_{z=z/z_o} = X(z/z_o)$$

In particular,

$$\left(e^{j\omega_o}\right)^n x[n] \Leftrightarrow X(z/e^{j\omega_o})$$

Rotation in z-plane

where

$$X\left(\frac{z}{e^{j\omega_o}}\right) = X\left(re^{j\omega}e^{-j\omega_o}\right) = X\left(re^{j(\omega-\omega_o)}\right)$$

This is z-domain version of frequency-shift property

Properties of z-transform

5. BIBO Stability:

BIBO stable **only if the ROC includes unit circle**,

since Fourier transform must exist

and Fourier transform lies on unit circle

6. Initial value:

$$\lim_{z\to\infty} X(z) = x[0] \quad \text{for causal x[n]}$$

$$\text{since} \quad \lim_{z\to\infty} X(z) = \lim_{z\to\infty} \sum_{n=0}^{\infty} x[n] z^{-n}$$

Inverse Z-Transform Methods

1) Contour integral $x[n]=\frac{1}{2\pi j}\oint X(z)z^{n-1}\,dz$
2) Lookup in the table and don't forget the ROC!! (will later cover partial fraction expansion)
3) Power series expansion (long division)

- Long division example inverse z-transform of right-sided X(z)

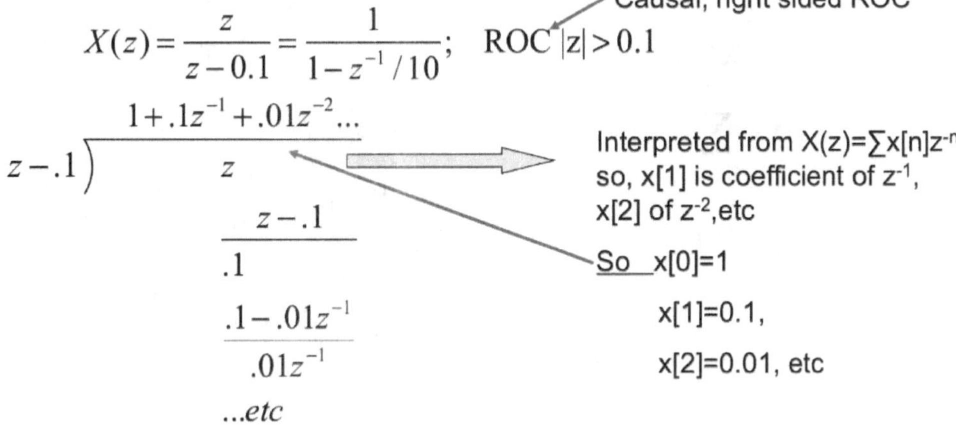

Causal, right sided ROC
$X(z)=\frac{z}{z-0.1}=\frac{1}{1-z^{-1}/10}$; ROC $|z|>0.1$
$1+.1z^{-1}+.01z^{-2}...$
$z-.1$) z
$z-.1$
$.1$
$.1-.01z^{-1}$
$.01z^{-1}$
...etc
Interpreted from $X(z)=\sum x[n]z^{-n}$
so, x[1] is coefficient of z^{-1},
x[2] of z^{-2},etc
So x[0]=1
x[1]=0.1,
x[2]=0.01, etc

Inverse z-transform

$$X(z)=\frac{z}{z-0.1}=\frac{1}{1-z^{-1}/10};\quad \text{ROC } |z|<0.1$$

- Example long-division inverse z-transform of left-sided X(z)
- Since ROC |z|<1, then use left sided solution below

$$\begin{array}{r|l} & -10z-100z^2... \\ -.1+z & z \\ & z-10z^2 \\ & 10z^2 \\ & 10z^2-100z^3 \\ & 100z^3 \\ & ...etc \end{array}$$

Change order

So:

x[-1]=-10

x[-2]=-100

etc

Significance of X(z) at z=1

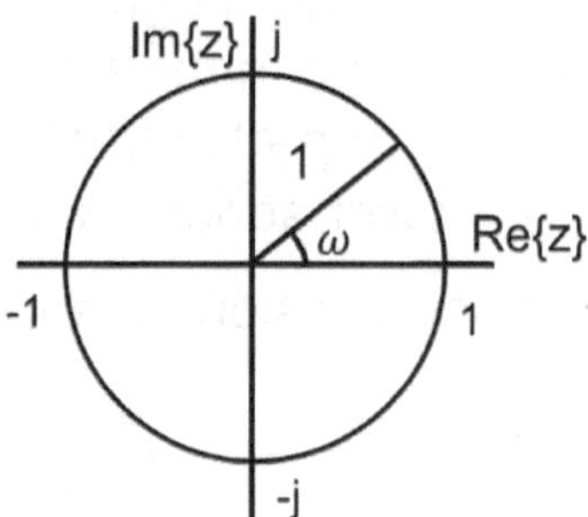

- X(z) at z=1 corresponds to dc
- Recall relation to DTFT

$$X(\omega) = X(z)\big|_{z=e^{j\omega}}$$

- It is simple to compute X(1):

$$X(1) = X(z)\big|_{z=1}$$

$$X(1) = \sum_{n=-\infty}^{\infty} x[n]z^{-n}$$

$$= \sum_{n=-\infty}^{\infty} x[n]$$

What DTFT frequency ω corresponds to z=1?

z=-1?

example:

$$X(z) = \frac{z}{z-0.1}; \quad \text{ROC } |z|>0.1$$

$$\text{X}(1) = \frac{1}{1-0.1} = \frac{1}{0.9}$$

DFT vs. Z-Transform

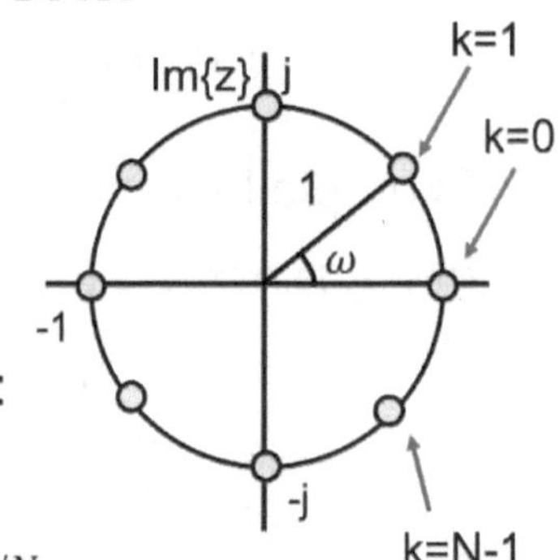

- X(z) on unit circle is X(ω)
- Recall relation to DTFT

$$X(\omega) = X(z)\big|_{z=e^{j\omega}}$$

- DFT corresponds to samples of DTFT:

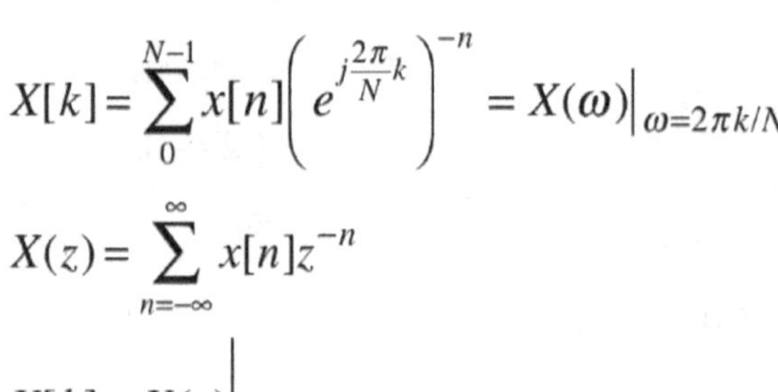

$$X[k] = \sum_{0}^{N-1} x[n]\left(e^{j\frac{2\pi}{N}k}\right)^{-n} = X(\omega)\big|_{\omega=2\pi k/N}$$

$$X(z) = \sum_{n=-\infty}^{\infty} x[n]z^{-n}$$

$$X[k] = X(z)\Big|_{z=e^{j\frac{2\pi}{N}k}}$$

What is N for DFT points shown above?

Example Inverse Z-Transform

- Example inverse z-transform of right-sided X(z)

$$X(z)=\frac{1-z^{-4}}{1-z^{-1}}=\frac{z^4-1}{z^4-z^3};\quad |z|>0$$ Causal ROC

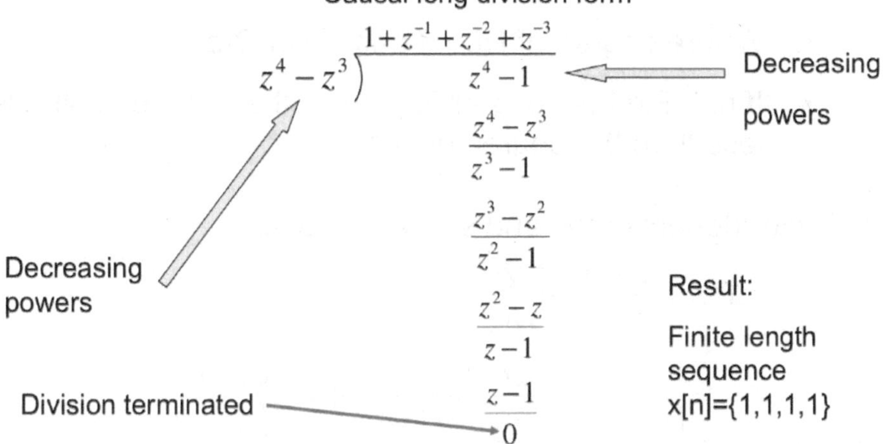

Example

- Same example, but using powers of z^{-1}

$$X(z)=\frac{1-z^{-4}}{1-z^{-1}}=\frac{z^4-1}{z^4-z^3};\quad |z|>0$$

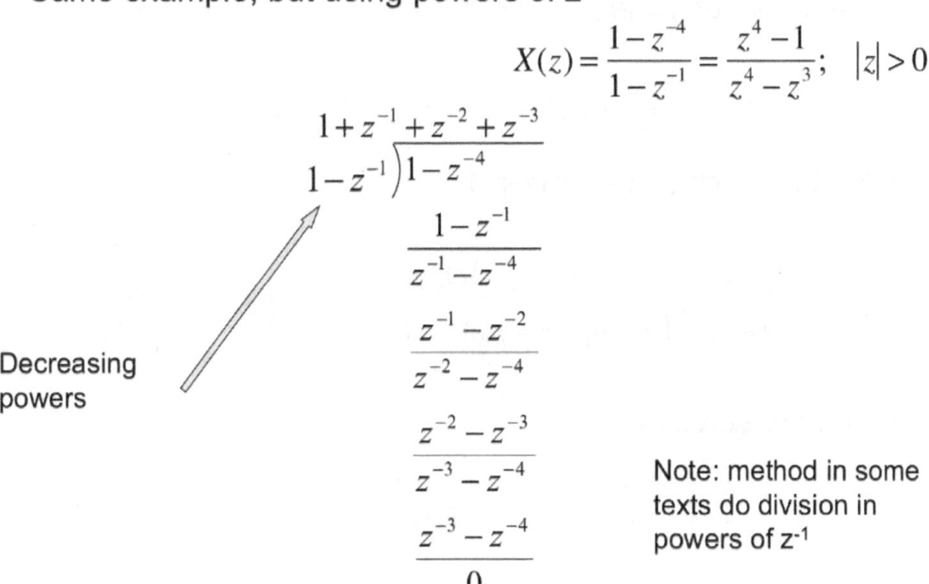

Inverse z-transform by Partial Fraction Expansion

- **Consider partial fraction expansion for first-order poles**

1) Suppose X(z)=N(z)/D(z),
   - Where degree of N(z) < degree of D(z)
   - If not: First do long division until the degree of N(z) is less than the degree of D(z)

2) Factor denominator, and set N(z) = zQ(z)

$$X(z)=\frac{N(z)}{D(z)}=\frac{z\,Q(z)}{D(z)}$$

$$=\frac{z\,Q(z)}{(z-p_1)(z-p_2)\circ\circ\circ(z-p_i)}=z\sum_{\alpha=1}^{i}\frac{c_\alpha}{z-p_\alpha}$$

Partial Fraction Expansion

From previous slide

$$\frac{Q(z)}{(z-p_1)(z-p_2)\circ\circ\circ(z-p_i)}=\sum_{\alpha=1}^{i}\frac{c_\alpha}{z-p_\alpha}$$

3. Multiply both sides by $(z-p_k)$

$$(z-p_k)\frac{Q(z)}{D(z)}=\frac{(z-p_k)Q(z)}{(z-p_1)(z-p_2)\circ\circ\circ(z-p_i)}=(z-p_k)\left\{\sum_{\alpha=1}^{i}\frac{c_\alpha}{z-p_\alpha}\right\}$$

4. At $z=p_k$ solve for c_k

$$(z-p_k)\frac{Q(z)}{D(z)}\bigg|_{z=p_k}=c_k$$

Partial Fraction Expansion

From previous slide

6. Having solved for c_k the partial fraction expansion is:

$$X(z)=\frac{N(z)}{D(z)}=z\frac{Q(z)}{D(z)}=z\sum_{\alpha=1}^{i}\frac{c_\alpha}{z-p_\alpha}=\sum_{\alpha=1}^{i}\frac{zc_\alpha}{z-p_\alpha}$$

7. Taking inverse z-transform

$$a^n u[n]\Leftrightarrow\frac{z}{z-a};\quad |z|>|a|$$

$$So:\quad x[n]=Z^{-1}\{X(z)\}=Z^{-1}\left\{\sum_{\alpha=1}^{i}\frac{z\,c_\alpha}{z-p_\alpha}\right\}=\sum_{\alpha=1}^{i}c_\alpha(p_\alpha)^n u[n]$$

Note: method in text does partial fraction expansion in powers of z^{-1}

Partial Fraction Expansion Example

- Partial Fraction Expansion Example:

$$X(z)=\frac{1}{(z-1/2)(z-1/4)}=z\frac{Q(z)}{D(z)}=z\frac{1}{(z-1/2)(z-1/4)z};\quad ROC\,|z|>1/2$$

$$=z\left(\frac{\left.\frac{(z-1/2)Q(z)}{D(z)}\right|_{z=1/2}}{z-1/2}+\frac{\left.\frac{(z-1/4)Q(z)}{D(z)}\right|_{z=1/4}}{z-1/4}+\frac{\left.\frac{(z)Q(z)}{D(z)}\right|_{z=0}}{z}\right)$$

$$=z\left(\frac{\left.\frac{1}{(z-1/4)z}\right|_{z=1/2}}{z-1/2}+\frac{\left.\frac{1}{(z-1/2)z}\right|_{z=1/4}}{z-1/4}+\frac{\left.\frac{1}{(z-1/4)(z-1/2)}\right|_{z=0}}{z}\right)=\frac{8z}{z-1/2}-\frac{16z}{z-1/4}+8$$

- ROC indicates causal right-sided x[n]
- No need for long division since numerator degree is less than denominator

Example, continued

- Rearrange partial fraction expansion results into forms where x[n] may be found using z-transform tables
- Use z-transform properties tables, as needed
- For preceding example:

$$= \frac{8z}{z-1/2} - \frac{16z}{z-1/4} + 8$$

$$so:$$

$$x[n] = 8\left(\frac{1}{2}\right)^n u[n] - 16\left(\frac{1}{4}\right)^n u[n] + 8\delta[n]$$

Example (powers of z^{-1} instead)

- Same problem, per method in text (using powers of z^{-1} instead of powers of z)

$$X(z) = \frac{1}{(z-1/2)(z-1/4)} = \frac{z^{-2}}{\left(1-\frac{1}{2}z^{-1}\right)\left(1-\frac{1}{4}z^{-1}\right)}$$

$$= \frac{z^{-2}}{1-\frac{3}{4}z^{-1}+\frac{1}{8}z^{-2}}$$

- First, denominator highest power of $z^{-1}=(z^{-1})^2$. So, denominator order =2
- Numerator order =2, since it has term $(z^{-1})^2$
- So, must first do long division

Example, continued

Long division

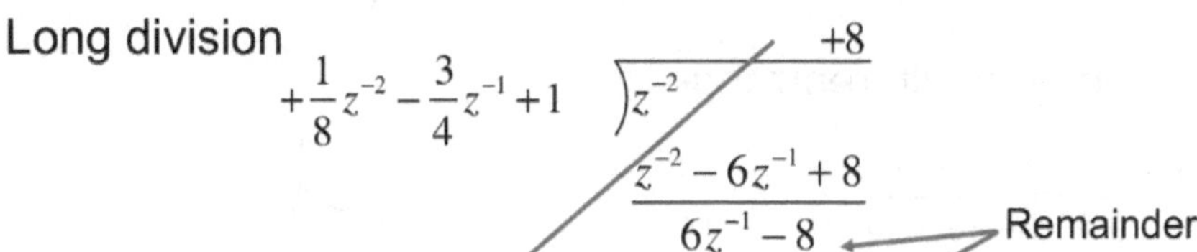

+8
$+\frac{1}{8}z^{-2}-\frac{3}{4}z^{-1}+1 \;)\; z^{-2}$
$z^{-2}-6z^{-1}+8$
$6z^{-1}-8$
Remainder

So, taking remainder, and doing expansion in powers of z^{-1} :

$$X(z)=8+\frac{-8+6z^{-1}}{\left(1-\frac{1}{2}z^{-1}\right)\left(1-\frac{1}{4}z^{-1}\right)}$$

$$=8+\frac{\left.\frac{-8+6z^{-1}}{1-(1/4)z^{-1}}\right|_{z^{-1}=2}}{1-(1/2)z^{-1}}+\frac{\left.\frac{-8+6z^{-1}}{1-(1/2)z^{-1}}\right|_{z^{-1}=4}}{1-(1/4)z^{-1}}$$

Same result as before!

$$=8+\frac{8}{1-\frac{1}{2}z^{-1}}-\frac{16}{1-\frac{1}{4}z^{-1}}$$

Finding H(z) from difference equation

x[n] → h[n] → y[n]

- Example: 4-point sum

$$y[n]=\sum_{k=0}^{3}x[n-k]=x[n]+x[n-1]+x[n-2]+x[n-3]$$

$$similarly: y[n-1]=x[n-1]+x[n-2]+x[n-3]+x[n-4]$$

$$so \quad y[n]-y[n-1]=x[n]-x[n-4]$$

$$or \quad y[n]=y[n-1]+x[n]-x[n-4]$$

- Take z-transform of both sides:

$$Y(z)-z^{-1}Y(z)=X(z)-z^{-4}X(z) \quad or \quad Y(z)\left(1-z^{-1}\right)=\left(1-z^{-4}\right)X(z)$$

$$Y(z)=\frac{1-z^{-4}}{1-z^{-1}}X(z)=H(z)X(z) \quad so \quad H(z)=\frac{1-z^{-4}}{1-z^{-1}}$$

- For causal solution ROC must be outside outermost pole.

Poles and Zeroes of H(z)

- From previous result, rearrange

$$H(z)=\frac{1-z^{-4}}{1-z^{-1}}=\frac{z^4-1}{z^3(z-1)}$$
$$=\frac{(z-1)(z-j)(z+1)(z+j)}{z^3(z-1)}$$
$$=\frac{(z-j)(z+1)(z+j)}{z^3};$$
$$ROC \quad z\neq 0$$

Note: Pole-zero cancellation

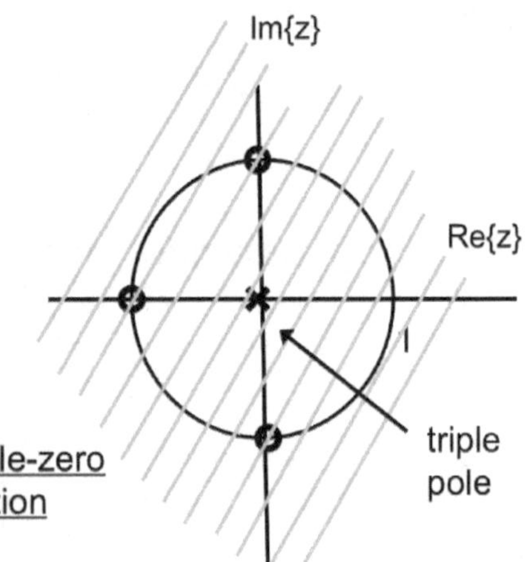

- Using long division show also:

$$H(z)=1+z^{-1}+z^{-2}+z^{-3}$$

Note: same poles and zeroes

Finding Difference Equation from H(z)

- Suppose H(z) below:

$$H(z)=\frac{z^4-1}{z^4-z^3}=\frac{1-z^{-4}}{1-z^{-1}}=\frac{Y(z)}{X(z)}$$
$$So:$$
$$\left(1-z^{-1}\right)Y(z)=\left(1-z^{-4}\right)X(z)$$
$$Y(z)-z^{-1}Y(z)=X(z)-z^{-4}X(z)$$

- Take inverse z-transform of both sides:

y[n] - y[n-1] = x[n] - x[n-4]

Example, continued

- Another form for the previous slide H(z) is:

$$H(z) = 1 + z^{-1} + z^{-2} + z^{-3}$$

- <u>so:</u>
  - y[n]=x[n]+x[n-1]+x[n-2]+x[n-3]

<u>Note</u>: H(z) can be factored in many ways to obtain a new form of the difference equation

what is h[n]?

Difference Equation and H(z) and ROC

Summary:

- H(z) can be determined directly from difference equation.
- Difference equation can be determined from H(z).
- Difference equation is not unique.
- For causal right-sided solution, ROC must be outside outermost pole.
- For left-sided solution, ROC must be inside innermost pole.
- BIBO stable ROC must include the unit circle.

Block Diagrams and H(z)

- Recall the z-transform of an exponential sequence

$$\frac{Y(z)}{X(z)} = H(z) = Z\left\{a^n u[n]\right\} = \frac{1}{1-az^{-1}} \quad |z| > |a|$$

- With difference equation:

  y[n] = x[n] + ay[n-1]

- Or z-transform of difference equation:

  $Y(z) - az^{-1}Y(z) = X(z)$

- A block diagram implementation of this system is shown

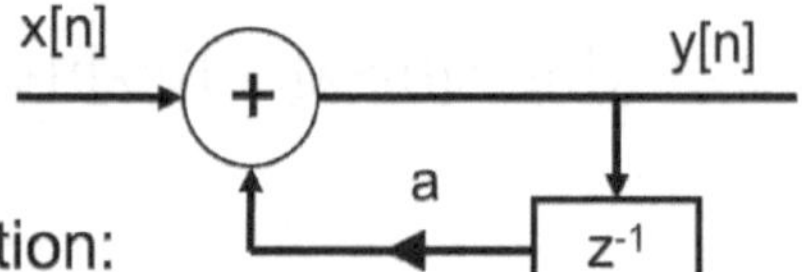

Block Diagram example

- Find the difference equation and H(z) for this block diagram

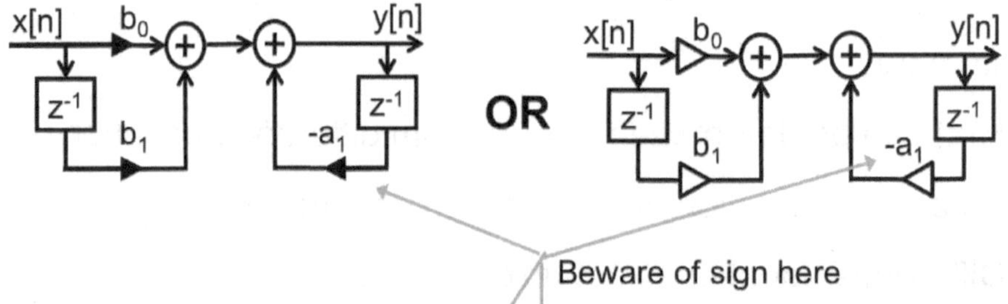

- Difference equation: $y[n]= -a_1 y[n-1]+b_0 x[n]+b_1 x[n-1]$
- Extensions of the above form yield general implementations of difference equations and H(z)
- For above example:

$$H(z) = \frac{Y(z)}{X(z)} = \frac{b_0 + b_1 z^{-1}}{1 + a_1 z^{-1}}$$

Inverse Systems

- Consider a signal X(z) distorted by a filter H(z)

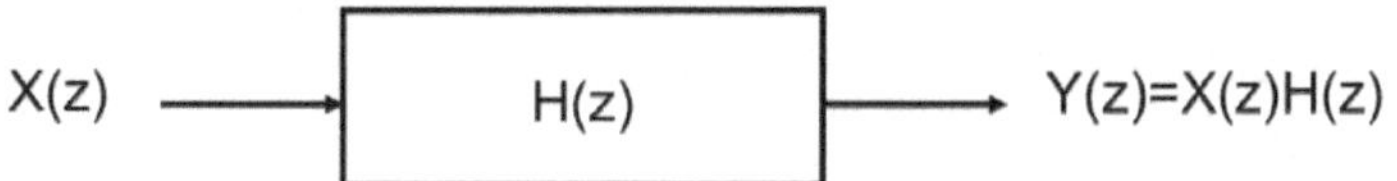

- Now, cascade this with inverse system $H_i(z)$

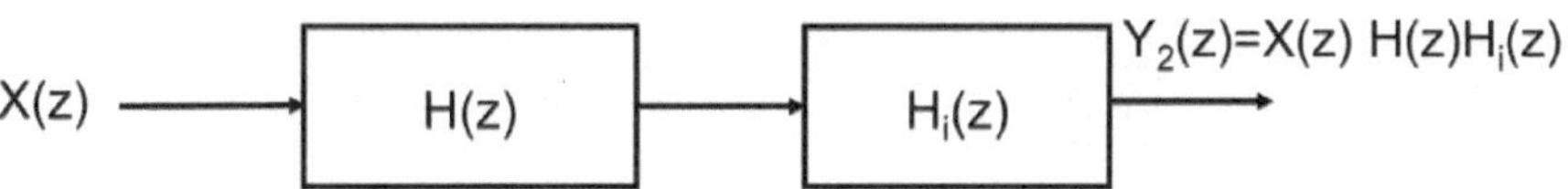

- So, to "undo" distortion:

$$\textit{if} \quad H_i(z) = \frac{1}{H(z)}$$

$$\textit{then} \quad Y_2(z) = X(z)H(z)H_i(z) = X(z)$$

$$\textit{so} \quad y_2[n] = x[n]$$

Note: ROC must be intersection of ROCs

So **poles and zeroes of H(z) must be inside unit circle** for poles of $H_i(z)$ to be inside unit circle and for both ROCs to include unit circle (causal stable)

Inverse System Examples

- Consider the possible inverses of the following system:

$$H(z) = \frac{1+\frac{4}{5}z^{-1}}{1-\frac{1}{2}z^{-1}}; \quad |z| > 1/2 \quad \text{(causal stable system)}$$

- Possible inverses are:

$$H_{i1}(z) = \frac{1-\frac{1}{2}z^{-1}}{1+\frac{4}{5}z^{-1}}; \quad |z| > 4/5$$ ← (causal stable inverse)

$$H_{i2}(z) = \frac{1-\frac{1}{2}z^{-1}}{1+\frac{4}{5}z^{-1}}; \quad |z| < 4/5$$ ← (Non-causal unstable inverse)

- Generally the causal stable inverse is used.
- Note: stable & causal Hi(z) may not be possible.

7 DIGITAL RADIO

The lecture notes in this chapter cover the application of digital signal processing in the implementation of digital radio systems.

Digital Radio Systems

- In theory, the "ideal digital radio" is simply an ADC (analog-to-digital converter), DAC (digital-to-analog converter), an antenna, and a signal processing system
- The entire radio is implemented in signal processing
- Therefore, some key radio concepts will be reviewed
- The discussed concepts apply to a broad range of systems such as cellphones, WiFi, cable television, and modems

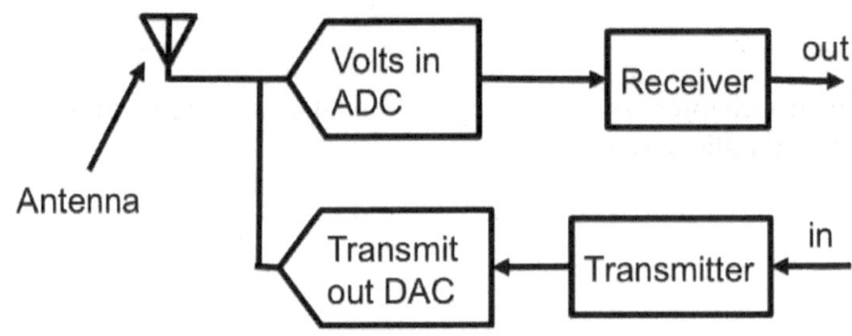

Frequency-shift property of the DTFT

- Frequency-shift property of the DTFT, as shown below:

$$f_{dsb-sc}[n] = m[n]\cos[\omega_c n], \quad where\ \mathrm{F}\ \{m[n]\} = M(\omega)$$

$$\mathrm{F}\ \{m[n]\cos[\omega_c n]\} = \sum_{n=-\infty}^{\infty} m[n]\cos[\omega_c n]e^{-j\omega n}$$

$$= \sum_{n=-\infty}^{\infty} m[n]\frac{1}{2}\left(e^{j\omega_c n} + e^{-j\omega_c n}\right)e^{-j\omega n} = \sum_{n=-\infty}^{\infty} m[n]\frac{1}{2}\left(e^{j(\omega_c-\omega)n} + e^{-j(\omega_c+\omega)n}\right)$$

$$= \frac{1}{2}\sum_{n=-\infty}^{\infty} m[n]e^{-j(-\omega_c+\omega)n} + \frac{1}{2}\sum_{n=-\infty}^{\infty} m[n]e^{-j(\omega_c+\omega)n}$$

$$= \frac{1}{2}\left[M(\omega-\omega_c) + M(\omega+\omega_c)\right]$$

So:

$$m[n]\cos[\omega_c n] \Leftrightarrow \frac{1}{2}\left[M(\omega-\omega_c) + M(\omega+\omega_c)\right]$$

Frequency shift property of DTFT

Note: DTFT is still periodic

M(ω)
A
ω
0

A/2
A/2
ω
$-\omega_c$
0
ω_c

Digital Radio DSB-SC Modulation

- The most common method to shift a baseband signal m[n] to high frequency is by using DSB-SC modulation (double-sideband suppressed-carrier modulation)
- In a digital radio, DSB-SC is implemented by multiplying the signal m[n] by a sine or cosine at the desired radio frequency
- Analysis for a discrete-time implementation uses the frequency-shift property of the DTFT, as shown below:

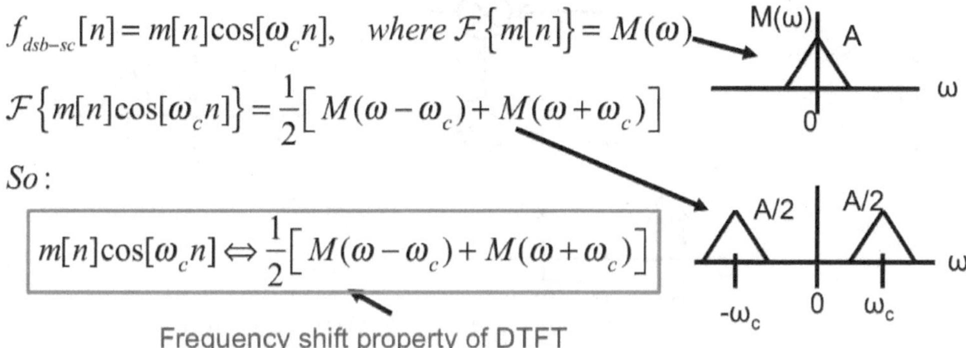

DSB-SC Transmitter Block Diagram

- Block diagram of DSB-SC system

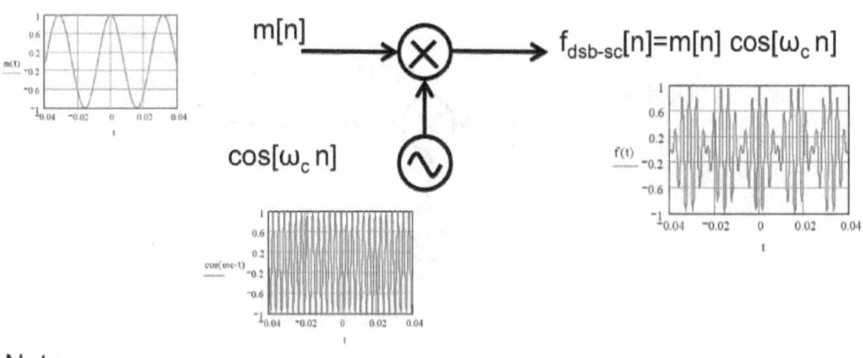

Note:

Although the transmitter is shown above as a "hardware block diagram," in a digital radio the functions may be implemented in software or digital hardware.

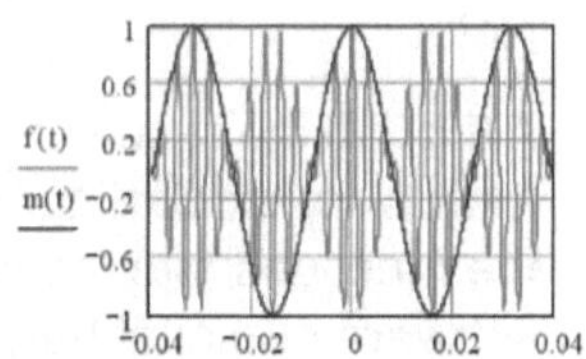

DSB-SC Transmitter, Frequency Domain

- Block diagram of DSB-SC transmitter system
- Recall: *all DTFT are periodic* with period 2π, even though not illustrated below

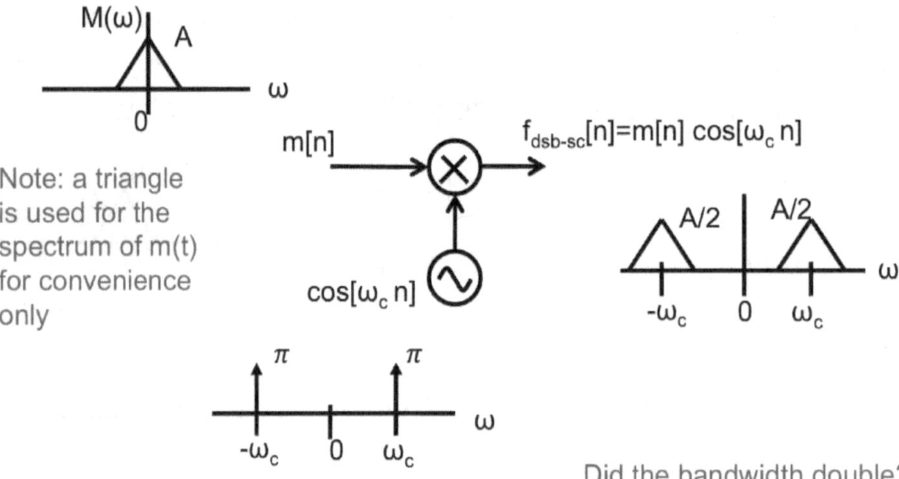

DSB-SC Receiver Block Diagram

- Block diagram of DSB-SC system
- What lowpass filter parameters give an output equal to m(t)?

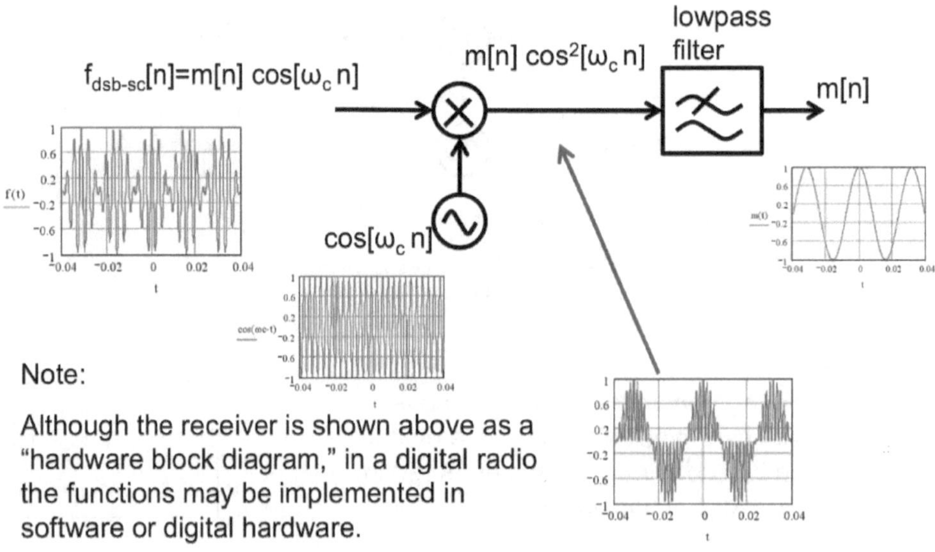

Note:

Although the receiver is shown above as a "hardware block diagram," in a digital radio the functions may be implemented in software or digital hardware.

DSB-SC Receiver Block Diagram

- Block diagram of DSB-SC system
- The lowpass filter should have the same bandwidth as m[n]

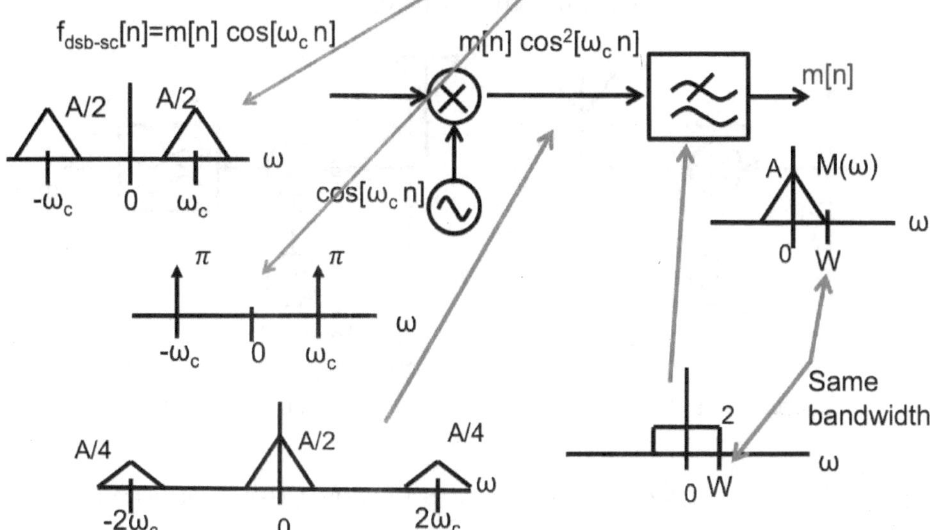

Bandwidth

- Recall definitions of bandwidth

- Ideal low-pass systems
  - See figure at right
  - Bandwidth is 0 to cutoff

- Ideal bandpass systems
  - See figure at right
  - Bandwidth is full width of positive frequencies

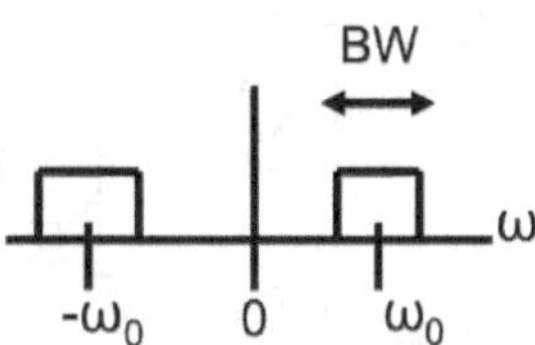

DSB-SC Transmitter/Receiver System

- Block diagram of DSB-SC Tx/Rx system

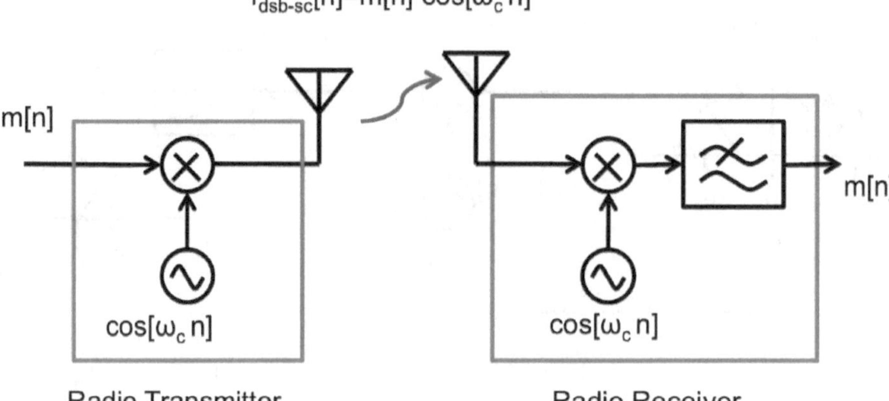

- At receiver, miles away, how is cosine phase determined?

Quadrature Amplitude Modulation

- QAM is used in most modern digital systems
- Takes advantage of orthogonality of sin() and cos()
- Results in "double-use" of frequency channel

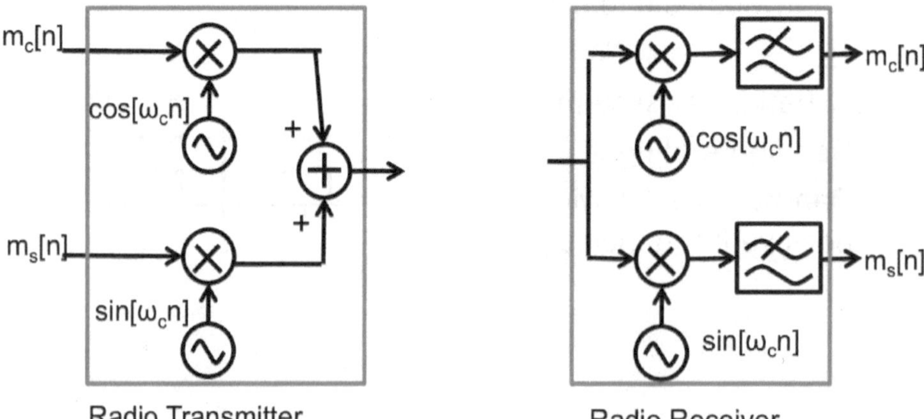

Quadrature Amplitude Modulation

- QAM can create amplitude and phase modulation
- Consider outputs when $m_c(t)=+/-1$ or $m_s(t)=+/-1$
- Constellation can be created from receiver output

$$f_{qam}[n]=m_c[n]\cos[\omega_c n]+m_s[n]\sin[\omega_c n]$$

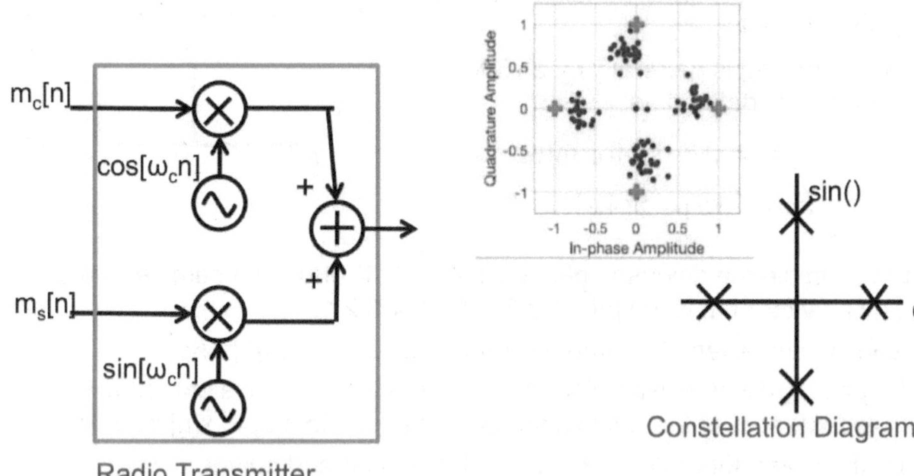

Quadrature Amplitude Modulation Theory

- The following is a summary of theory of the QAM transmitter and receiver

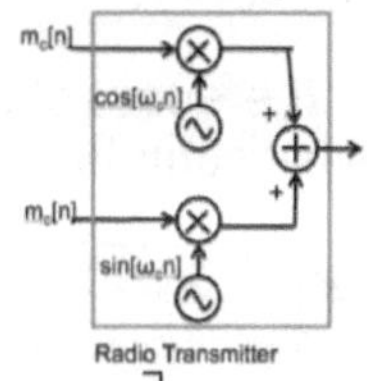

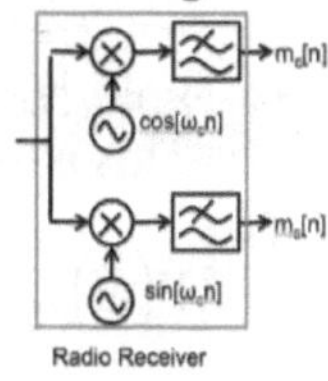

$$f_{qam}[n]=m_c[n]\cos[\omega_c n]+m_s[n]\sin[\omega_c n]$$

$$\mathcal{F}\{m_c[n]\cos[\omega_c n]\}=0.5\left[M_c(\omega-\omega_c)+M_c(\omega+\omega_c)\right]$$

$$\mathcal{F}\{m_s[n]\sin[\omega_c n]\}=-0.5j\left[M_s(\omega-\omega_c)-M_s(\omega+\omega_c)\right]$$

receiver cosine channel output:

$$\cos[\omega_c n]\left(m_c[n]\cos[\omega_c n]+m_s[n]\sin[\omega_c n]\right)$$

$$=0.5\left\{m_c[n]\left(1+\cos[2\omega_c n]\right)+m_s[n]\sin[2\omega_c n]\right\}$$

and after lowpass filter, receiver output $=m_c[n]$

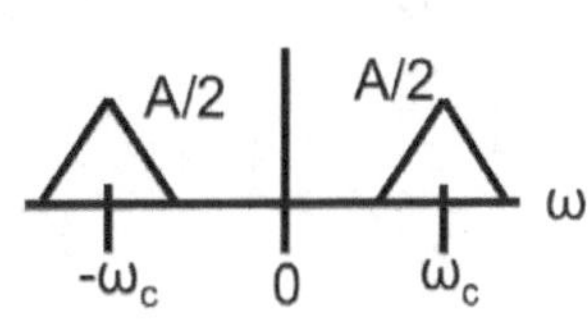

receiver sine channel output:

$$\sin[\omega_c n]\left(m_c[n]\cos[\omega_c n]+m_s[n]\sin[\omega_c n]\right)$$

$$=0.5\left\{m_c[n]\sin[2\omega_c n]+m_s[n]\left(1-\cos[2\omega_c n]\right)\right\}$$

and after lowpass filter, receiver output $=m_s[n]$

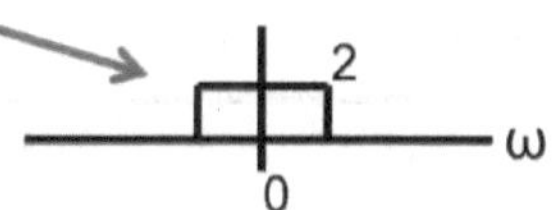

Downsampling

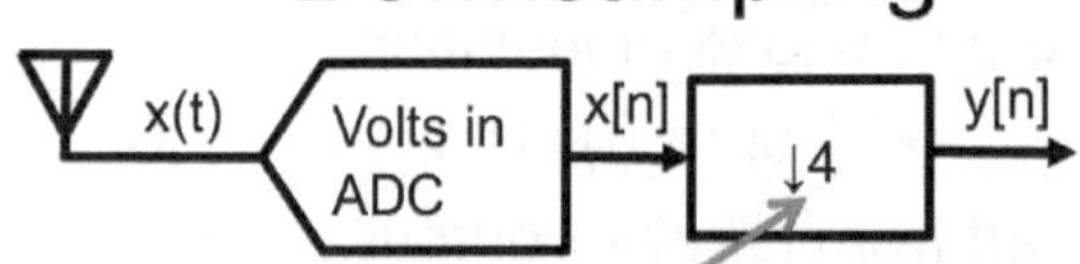

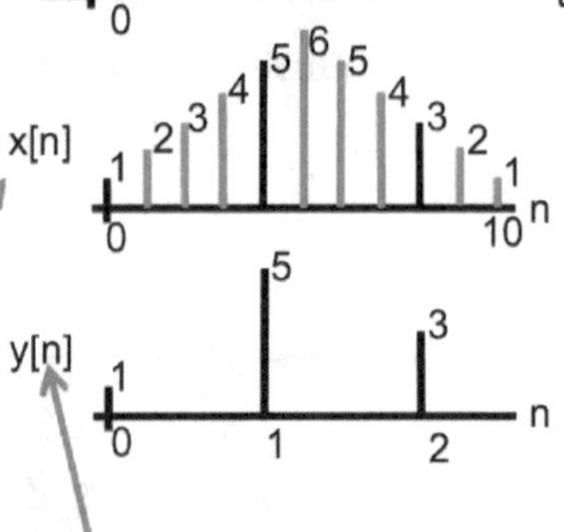

- In the example above the downsampler is denoted "↓4" and downsamples signal x[n] by a factor of 4
- Downsampling is generally by an integer factor, and is defined as:

$$y[n] = x[kn] \quad \text{for integer } k$$

$$here: y[n] = x[4n]$$

- In the example the downsampler is denoted "↓4" and downsamples signal x[n] by a factor of 4, with x[n]={1,2,3,4,5,6,5,4,3,2,1,...}
- As shown, only every 4th sample is retained, so with y[n]={1,5,3,...}
- Note: after x[n] is downsampled, the resulting y[n] is equivalent to having sampled x(t) at ¼ of the original sample rate used to obtain x[n] from xt)
- Note: the lower data rate saves computation time and power

Downsampling Effect on Spectrum

- Next, consider the effect of downsampling on the spectrum
  This is somewhat complicated to derive, so it is just given here:

$$Y(\omega) = \frac{1}{k}\sum_{\alpha=0}^{k-1} X\left(\frac{\omega}{k} - \frac{2\pi\alpha}{k}\right) \qquad \text{where DTFT is: } X(\omega) = \sum_{n=-\infty}^{\infty} x[n] z^{-n}\Big|_{z=e^{j\omega}}$$

- Note: Y(ω) =2π corresponds to1/k times original sample frequency in Hz, so continuous time sampled spectrum $Y_s(\Omega)$ aliases at ¼ the original rate
- And the $Y_s(\Omega)$ amplitude is ¼ of $X_s(\Omega)$

x(t), t, 0; x[n], 0, 1, 10, n; y[n], 0, 1, 2, n

DTFT spectra: X(ω), A/T_s, -2π, 2π, ω; Y(ω), A/(4T_s), -2π, 2π, ω

Continuous spectra: X(Ω), A, 0, Ω; $X_s(\Omega)$, A/T_s, -Ω_s, Ω_s, Ω; $Y_s(\Omega)$, A/(4T_s), -Ω_s/4, Ω_s/4, Ω

1/4 original sample rate

8 DIGITAL FILTER DESIGN

The lecture notes in this chapter cover classical digital filter design methods, including impulse invariance, bilinear transform, and windowing methods.

Digital Filters

Note on Nyquist Limit and Subsampling

- Nyquist, better stated, is that a signal must be sampled at twice the bandwidth, instead of twice the highest frequency
- For a bandpass signal this may be much lower sampling rate than "twice the highest frequency"
- See various papers on subsampling/undersampling
- This is an important concept used in many radio designs

- However: For purposes of this course, assume that a signal must be sampled at twice the highest frequency, *unless explicitly stated otherwise*

Continuous-Time Filtering with DSP

- DSP is often used to replace some analog system
- The analog system is defined by its frequency response $\mathbf{H_a(\Omega)}$
- The effective frequency response of the DSP system is $\mathbf{H_{eff}(\Omega)}$
- The goal is then to design the DSP so that $\mathbf{H_{eff}(\Omega) = H_a(\Omega)}$

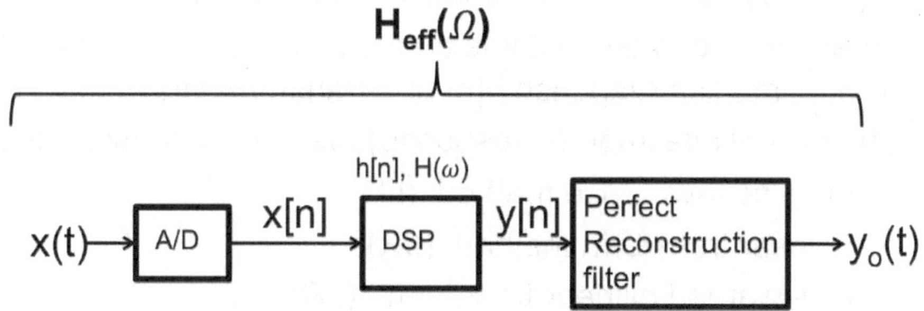

- Note: "filter design" $H(\omega)$ or h[n] can be any LTI system!! **...not "just filters" such as Butterworth, etc.**

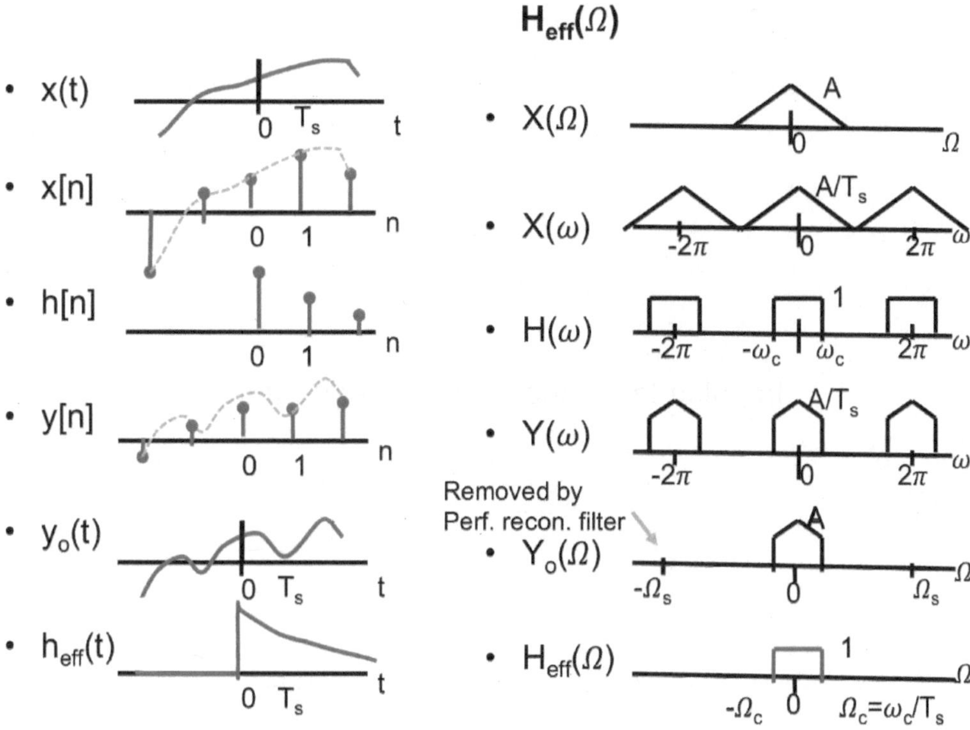

Filter Design Methods

- The following sections discuss four filter design methods
- Generally, the four methods start with some analog continuous-time filter design
- The continuous-time filter is defined by impulse response $h_c(t)$, or by continuous-time Laplace transform $H_c(s)$
- The continuous-time filter is then converted to a digital filter having impulse response h[n] or z-transform H(z)
- The IIR (infinite impulse response) methods discussed are:
    - Impulse invariance method, (IIR)
    - Bilinear transform method, (IIR)
    - Pre–warped bilinear transform, (IIR)
- The FIR (finite impulse response) methods discussed are:
    - Windowing methods, (FIR)

Impulse Invariance Filter Design Method

Impulse Invariance Method: In Time Domain

- The impulse invariance design method can be implemented in the time domain as:
  - $h[n] = T_s\, h_c(nT_s)$
  - i.e., sample the continuous impulse response at T_s

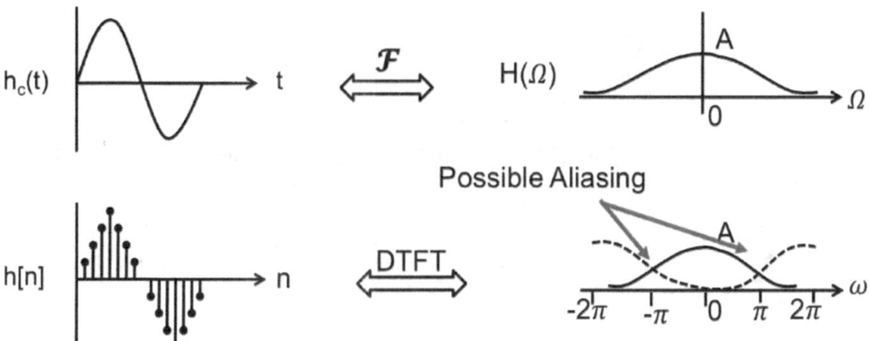

Impulse Invariance Method: In Frequency Domain

- The impulse invariance design method can also be implemented in the frequency domain
  - In this, the digital filter is based on the Laplace transform representation the analog filters where, $L\{h_c(t)\} = H_c(s)$
- Derivation of the method
  1. Given the continuous-time filter, with $H_c(s)$ of the form

$$H_c(s) = \sum_{k=1}^{N} \frac{A_k}{s - s_k}$$

  2. First find $h_c(t)$, and $h[n]$ using time-domain method:

$$h_c(t) = L^{-1}\{H_c(s)\} = L^{-1}\left\{\sum_{k=1}^{N} \frac{A_k}{s - s_k}\right\} = \sum_{k=1}^{N} A_k e^{s_k t} u(t)$$

$$then: \quad h[n] = T_s\, h_c(nT_s) = T_s \sum_{k=1}^{N} A_k e^{s_k n T_s} u[n]$$

Impulse Invariance, Frequency Domain Method

- Continuing,
  3. Taking z-transform of both sides:

$$H(z) = Z\{h[n]\} = Z\left\{T_s \sum_{k=1}^{N} A_k e^{s_k n T_s} u[n]\right\}$$

$$= \sum_{k=1}^{N} \frac{T_s A_k}{1 - e^{s_k T_s} z^{-1}}$$

- In summary, the impulse invariance in frequency-domain is:

$$H_c(s) = \sum_{k=1}^{N} \frac{A_k}{s - s_k} \quad \Rightarrow \quad H(z) = \sum_{k=1}^{N} \frac{T_s A_k}{1 - e^{s_k T_s} z^{-1}}; \quad |z| > \text{outermost pole}$$

Impulse Invariance Method: Summary

- Impulse invariance filter design can be done in time domain or frequency domain
  - o Both approaches suffer from aliasing
  - o Both approaches are equivalent
- Impulse invariance method in time domain:

$$h[n] = T_s\, h_c(n\, T_s)$$

- Impulse invariance method in frequency domain:

$$H_c(s) = \sum_{k=1}^{N} \frac{A_k}{s - s_k} \quad \Rightarrow \quad H(z) = \sum_{k=1}^{N} \frac{T_s A_k}{1 - e^{s_k T_s} z^{-1}}; \quad |z| > \text{outermost pole}$$

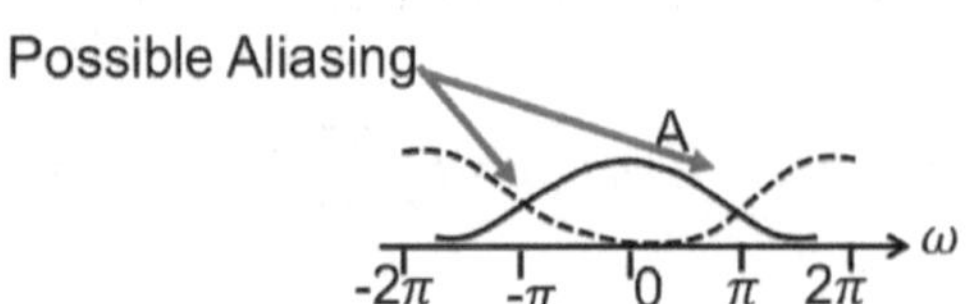

Impulse Invariance Filter Stability

- Important aspects of Impulse Invariance
  - Time and frequency domain methods are equivalent
  - Aliasing is an issue
  - A stable analog filter guarantees stable H(z)
- Proof of stability property:

$$H(z)=\sum_{k=1}^{N}\frac{A_k T_s}{1-e^{s_k T_s}z^{-1}} \quad \text{has poles at } z=e^{s_k T_s}$$

$$if \quad s=\sigma+j\Omega,$$

$$\text{then } e^{s_k T_s}=e^{\sigma T_s+j\Omega T_s}=e^{\sigma T_s}e^{j\Omega T_s}$$

$$and\left|e^{\sigma T_s}e^{j\Omega T_s}\right|<1 \quad for\ \sigma<0\ (left-plane\ poles)$$

- So, stable poles in $H_c(s)$ assure stable poles for H(z)

Example: 2nd order Butterworth

- Impulse Invariance design from $H_c(s)$

$$H_c(s)=\frac{1}{\left(s/\Omega_c-e^{j3\pi/4}\right)\left(s/\Omega_c-e^{-j3\pi/4}\right)}=\frac{\Omega_c^2}{s^2+s\Omega_c\sqrt{2}+\Omega_c^2}$$

$$H_c(s)=\Omega_c\left(\frac{j/\sqrt{2}}{\left(s-\Omega_c e^{j3\pi/4}\right)}-\frac{j/\sqrt{2}}{\left(s-\Omega_c e^{-j3\pi/4}\right)}\right)$$

$$H(z)=T_s\Omega_c\left(\frac{j/\sqrt{2}}{\left(1-e^{\Omega_c T_s e^{j3\pi/4}}z^{-1}\right)}-\frac{j/\sqrt{2}}{\left(1-e^{\Omega_c T_s e^{-j3\pi/4}}z^{-1}\right)}\right)$$

Since, $\frac{A_k}{s-s_k} \Rightarrow \frac{A_k T_s}{1-e^{s_k T_s}z^{-1}}$

Bilinear Transform Filter Design Method

Bilinear Transform Filter Design Method

- Impulse invariance method is susceptible to aliasing issues
- The bilinear transform filter design method avoids aliasing by "squeezing" $-\infty < \Omega < \infty$ into $-\pi < \omega < \pi$
- The bilinear transform is a frequency-domain method
- Given continuous-time filter $H_c(s)$, the bilinear transform yields:

$$H(z) = H_c(s)\Big|_{s=\frac{2}{T_s}\frac{z-1}{z+1}=\frac{2}{T_s}\frac{1-z^{-1}}{1+z^{-1}}}$$

- Thus, simply replace "s" in $H_c(s)$ by $(2/T_s)(z-1)/(z+1)$
- Important aspects of bilinear transform
  - No aliasing
  - Stable continuous-time filter guarantees stable H(z)
  - Results in a frequency-warped filter response
- Since the $-\infty < \Omega < \infty$ continuous frequency maps into $-\pi < \omega < \pi$ discrete-time frequency, it must be "warped" or "squeezed"

Bilinear Transform Stability

- Let $s = \sigma + j\Omega$ and $z = r e^{j\omega}$
- Then the bilinear transform substitution becomes:
  $s = (2/T_s)(z-1)/(z+1)$, which implies that
  $\sigma + j\Omega = (2/T_s)(re^{j\omega} - 1)/(re^{j\omega} + 1)$
- After much rearrangement, solving for σ and Ω:

$$\sigma = \frac{2}{T_s}\frac{r^2-1}{1+r^2+2r\cos(\omega)} \qquad \Omega = \frac{2}{T_s}\frac{2r\sin(\omega)}{1+r^2+2r\cos(\omega)}$$

- For both, denominator minimum = $r^2+1-2r = (1-r)^2$
- So, denominator is always positive
- And, r>1 ensures that $\sigma>0$, and r<1 assures $\sigma<0$
- Therefore, stable poles in $H_c(s)$ assure stable poles for H(z)

Bilinear Transform, Frequency Warping

- Of particular interest is the unit circle in z-plane where r=1, then $\sigma=0$, and then solving for Ω

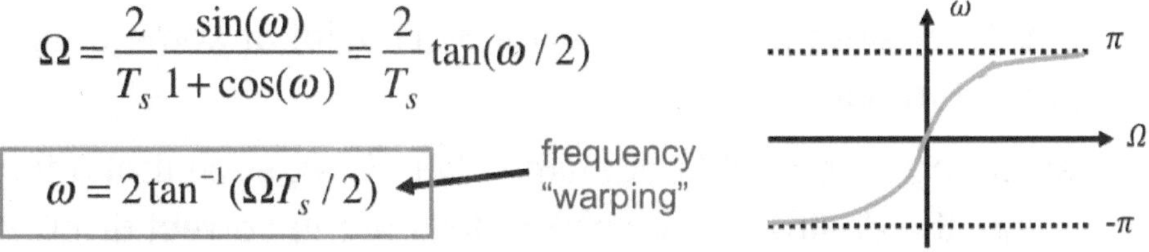

$\Omega = \frac{2}{T_s}\frac{\sin(\omega)}{1+\cos(\omega)} = \frac{2}{T_s}\tan(\omega/2)$
$\omega = 2\tan^{-1}(\Omega T_s/2)$
frequency
"warping"
ω
π
Ω
-π

- This is the "Warping" found in Bilinear Transform
- Note: for small ω only, the relationship can be approximated as:

$$\Omega = \frac{2}{T_s}\frac{\sin(\omega/2)}{\cos(\omega/2)} \approx \frac{2}{T_s}\sin(\omega/2) \approx \frac{2}{T_s}(\omega/2) \approx \frac{\omega}{T_s}$$

Approximates ideal sampler relation: $\omega = \Omega T_s$

Pre-Warped Bilinear Transform

- Adjusting $H_c(s)$ can *correct* for warping at <u>one</u> frequency
- Consider an N-th order Butterworth: $|H_c(\Omega)|^2 = 1/(1+(\Omega/\Omega_c)^{2N})$, where Ω_c is 3-dB frequency in rad/s
- If designed with bilinear transform, where $\Omega = (2/T_s)\tan(\omega/2)$, then the frequency response of the digital filter would be:

$$|H(\omega)|^2 = |H_c(\Omega)|^2\Big|_{\Omega=(2/T_s)\tan(\omega/2)} = \frac{1}{1+\left(\frac{(2/T_s)\tan(\omega/2)}{\Omega_c}\right)^{2N}}$$

- Where the 3 dB frequency is $2\tan^{-1}(\Omega_c T_s/2)$ instead of $\Omega_c T_s$
- So, correct for this one frequency by changing the cutoff of the continuous-time analog filter to $\Omega'_c = (2/T_s)\tan(\Omega_c T_s/2)$
- Then, "pre-warped design" generates the proper 3 dB point

$$|H(\omega)|^2 = \frac{1}{1+\left(\frac{(2/T_s)\tan(\omega/2)}{(2/T_s)\tan(\Omega_c T_s/2)}\right)^{2N}}$$

3 dB point is at $\omega = \Omega_c T_s$

Pre-Warped Bilinear Transform: Summary

- The bilinear transform warps the frequency response
- Pre-warping can correct for frequency warping at <u>one</u> desired frequency, Ω_d
- Typically, desired frequency is the 3 dB point or some other important frequency
- First, change the continuous-time filter $H_c(s)$, such that after warping, the desired frequency Ω_d falls at the correct discrete-time frequency $\omega = \Omega_d T_s$
  - Typically, the continuous time filter is modified, such as shifting from Ω_d to $\Omega'_d = (2/T_s)\tan(\Omega_d T_s/2)$
- Finally, perform the bilinear transform on the pre-warped continuous time filter to obtain the discrete-time filter H(z)

Bilinear Transform: Butterworth Filter Example

- The continuous-time Butterworth filter is defined as:

$$|H_c(s)|^2 = \frac{1}{1+(s/j\Omega_c)^{2N}}$$

Where:

- N = order of the filter
- 1/2 power point is at s = $j\Omega_c$

$$H_c(s) = \frac{\Omega_c^2}{s^2 + s\Omega_c\sqrt{2} + \Omega_c^2}; \quad \text{for N=2, second order}$$

- The bilinear transform filter design is then:

$$H(z) = \left.\frac{\Omega_c^2}{s^2 + s\Omega_c\sqrt{2} + \Omega_c^2}\right|_{s=\frac{2}{T_s}\frac{z-1}{z+1}} = \frac{\Omega_c^2}{\left(\frac{2}{T_s}\frac{z-1}{z+1}\right)^2 + \Omega_c\sqrt{2}\left(\frac{2}{T_s}\frac{z-1}{z+1}\right) + \Omega_c^2}$$

$$= \frac{\Omega_c^2 T_s^2 (z+1)^2}{4(z-1)^2 + 2T_s\Omega_c\sqrt{2}(z-1)(z+1) + T_s^2\Omega_c^2(z+1)^2}$$

$$= \frac{\Omega_c^2 T_s^2 (z+1)^2}{\left(4 + 2\sqrt{2}T_s\Omega_c + T_s^2\Omega_c^2\right)z^2 + \left(2T_s^2\Omega_c^2 - 8\right)z + \left(4 - 2\sqrt{2}T_s\Omega_c + T_s^2\Omega_c^2\right)}$$

Pre-warped Bilinear Transform: Butterworth Example

- Consider the pre-warped bilinear transform design for second-order Butterworth filter parameters as follows:
  - Suppose f_s=8000 sample/s, T_s=1/8000, $\Omega_c = 2\pi\times 1000$
  - Then, $\Omega_c' = (2/T_s)\tan(\Omega_c T_s/2) = 2\pi \times 1055$, so
- The prewarped continuous-time filter becomes

$$|H(s)|^2 = \frac{1}{1 + \frac{s^4}{\left(j\frac{2}{T_s}\tan\left(\frac{\Omega_c T_s}{2}\right)\right)^4}} = \frac{1}{1+\left(\frac{s}{j(1055)2\pi}\right)^4}$$

3 dB pre-warped from 1000 to 1055 Hz

- The pre-warped bilinear transform filter design is then:

$$H_c(s) = \frac{\Omega_c^2}{s^2 + s\Omega_c\sqrt{2} + \Omega_c^2} = \frac{(2\pi\cdot 1055)^2}{s^2 + s(2\pi\cdot 1055)\sqrt{2} + (2\pi\cdot 1055)^2};$$

$$so: \quad H(z) = \left.\frac{(2\pi\cdot 1055)^2}{s^2 + s(2\pi\cdot 1055)\sqrt{2} + (2\pi\cdot 1055)^2}\right|_{s=\frac{2}{T_s}\frac{z-1}{z+1}}$$

Finding Butterworth Poles

$$|H_c(s)|^2 = \frac{1}{1+(s/j\Omega_c)^{2N}}$$

- Poles are at : $(s/j\Omega_c)^{2N} = -1$
- So,

$$(s/\Omega_c)^{2N} = -(j)^{2N} = -(-1)^N = \begin{cases} 1 & N\,odd \\ -1 & N\,even \end{cases}$$

Then: for N odd, poles are at: $s/\Omega_c = \sqrt[2N]{1}$ ← 2N roots of 1 on unit **circle**

for N even, poles are at: $s/\Omega_c = \sqrt[2N]{-1}$ ← 2N roots of -1 on unit **circle**

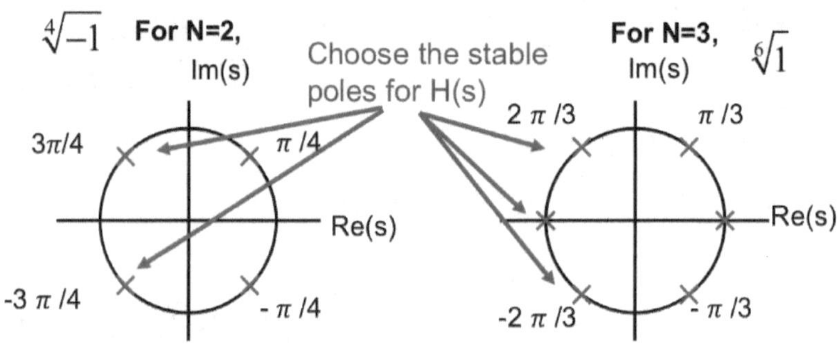

Windowing Filter Design Method

FIR Design by Windowing

- Assume a desired impulse response $h_d[n]$ is to be approximated by an M-point FIR approximation $h[n]$
- Windowing design method
  1. If needed, add delay to $h_d[n]$ to get approx. causal $h_d[n]$
  2. Then truncate $h_d[n]$ to M points by multiplying by a window function $w[n]$, as illustrated by rectangular window below
  3. So: $h[n]=h_d[n]\, w[n]$

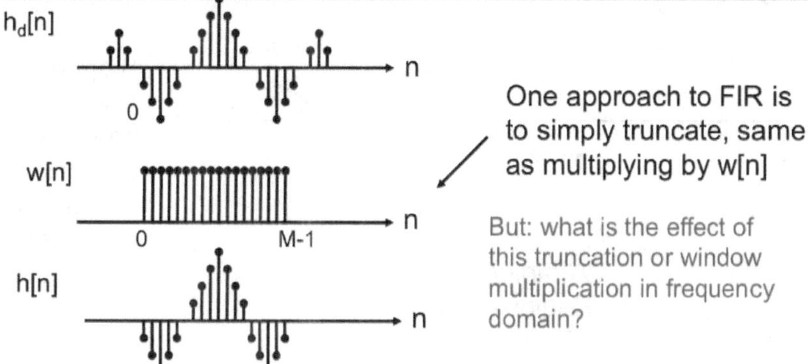

FIR Design by Windowing

- The effect in the frequency domain is periodic convolution

$$DTFT\{x[n]w[n]\} = \frac{1}{2\pi}\int_{-\pi}^{\pi} X(\beta)W(\omega-\beta)\,d\beta = \frac{1}{2\pi}\int_{-\infty}^{\infty}\left\{\Pi\left(\frac{\beta}{2\pi}\right)X(\beta)\right\}W(\omega-\beta)\,d\beta$$

- And, the resulting filter is the periodic convolution of the desired response $H_d(\omega)$ with the DTFT of the window $w[n]$
- For rectangular window, $W(\omega) = \sin(\omega M/2)e^{-j\omega(m-1)/2}/\sin(\omega/2)$
- Result is "smeared" transition and passband & stopband ripple

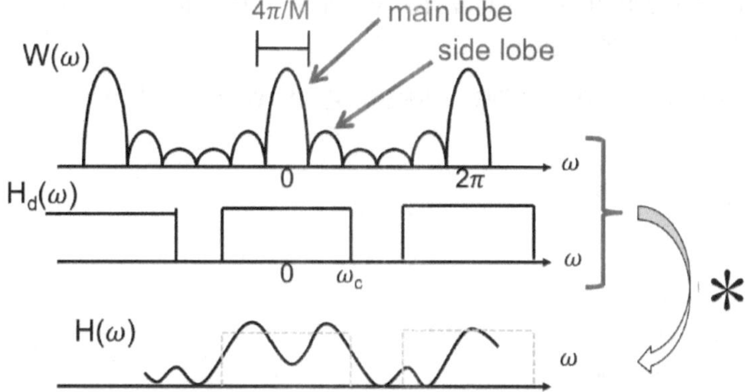

FIR Design by Windowing

Common windows of width M:

	Main Lobe	Peak Sidelobe
Rectangular	$4\pi/M$	-13 dB
Triangular(Bartlett)	$8\pi/M$	-26 dB
Hamming	$8\pi/M$	-43 dB

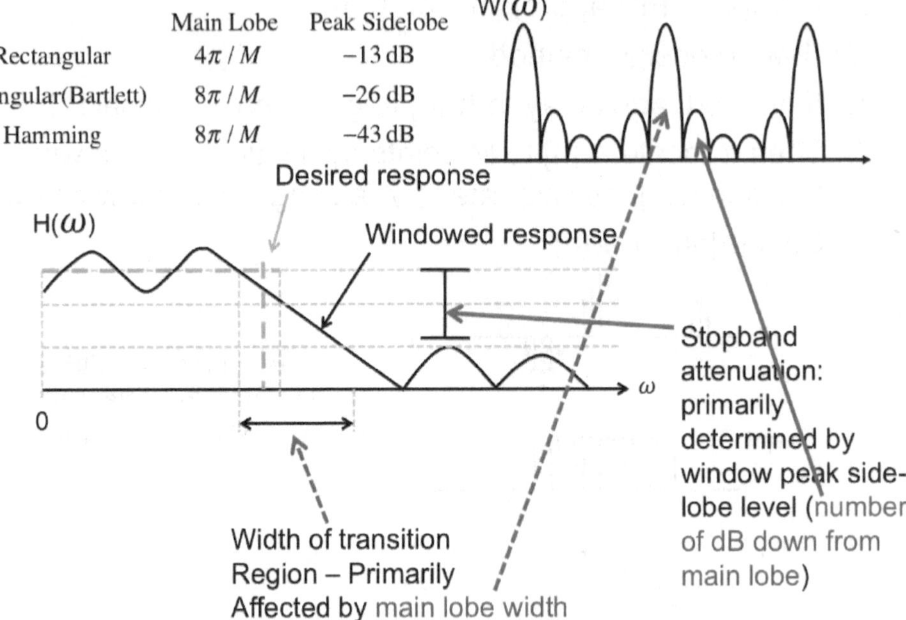

FIR Design by Windowing

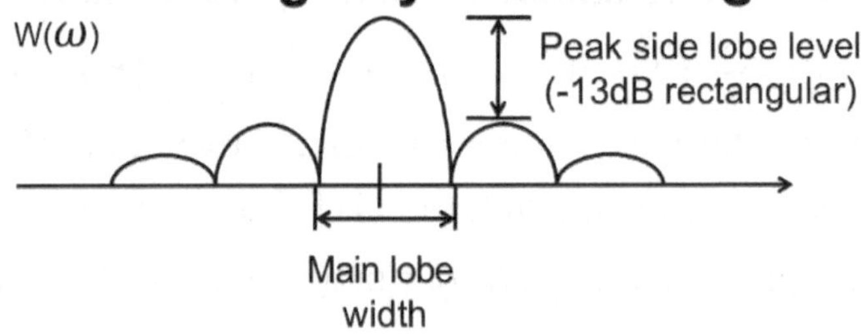

- So, choice of different window functions w[n] leads to different SPECTRAL tradeoffs.
  - Generally, wider mail lobe associated with a lower "side lobe"
- Other windows, see text

	Main Lobe	Peak Sidelobe
Rectangular	$4\pi/M$	-13 dB
Triangular(Bartlett)	$8\pi/M$	-26 dB
Hamming	$8\pi/M$	-43 dB

Filter Architecture

Filter Architecture

- As in earlier examples, the starting point for deriving digital filter architecture and block diagrams is the difference equation
- Although difference equations are not unique, some direct filter design forms are useful
- The general starting point is

Beware of sign difference in various books

$$H(z)=\frac{Y(z)}{X(z)}=\frac{\sum_{k'=0}^{M} b_{k'} z^{-k'}}{1+\sum_{k=1}^{N} a_k z^{-k}}=\frac{N(z)}{D(z)}$$

- Or, rearranging in difference equation form:

$$Y[z]=-a_1 z^{-1}Y(z)-a_2 z^{-2}Y(z)...+b_0 X(z)+b_1 z^{-1}X(z)+...$$

$$y[n]=-a_1 y[n-1]-a_2 y[n-2]...+b_0 x[n]+b_1 x[n-1]+...$$

Block Diagrams and Difference Equations

- Below left is the system block diagram for H(z)=Y(z)/X(z)
- Can rearrange block diagram differently to obtain a new form

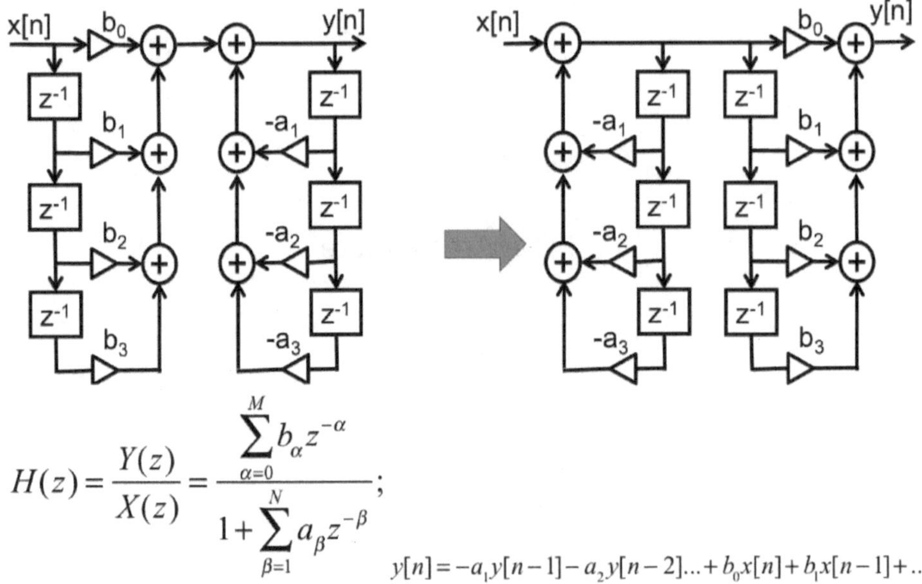

$$H(z)=\frac{Y(z)}{X(z)}=\frac{\sum_{\alpha=0}^{M} b_\alpha z^{-\alpha}}{1+\sum_{\beta=1}^{N} a_\beta z^{-\beta}};$$

$$y[n]=-a_1 y[n-1]-a_2 y[n-2]...+b_0 x[n]+b_1 x[n-1]+...$$

Block Diagrams and Difference Equations

- Merge registers, and the form on right requires less hardware

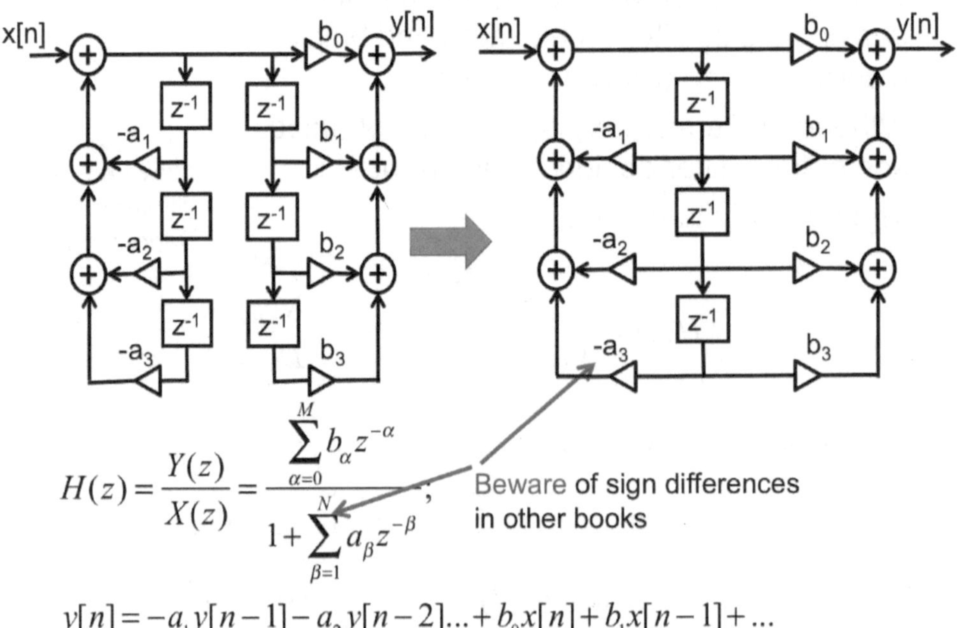

$$H(z)=\frac{Y(z)}{X(z)}=\frac{\sum_{\alpha=0}^{M} b_\alpha z^{-\alpha}}{1+\sum_{\beta=1}^{N} a_\beta z^{-\beta}};$$

$$y[n]=-a_1 y[n-1]-a_2 y[n-2]...+b_0 x[n]+b_1 x[n-1]+...$$

Summary of DSP Filter Design

- The foregoing filter design methods provide H(z)
- The filter architecture can be used to implement H(z) in hardware
- The difference equations can be used to implement the filter designs in software
- The effective DSP system freq. response is $\mathbf{H_{eff}(\Omega) = H_a(\Omega)}$

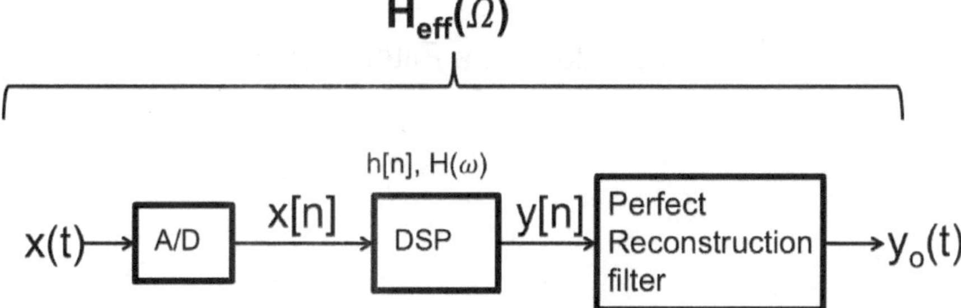

- Note: “filter design” H(ω) or h[n] can be any LTI system!! **...not “just filters” such as Butterworth, etc.**

Miscellaneous Filter Topics

Frequency Response from Poles & Zeroes

- Frequency response of a filter can be graphically estimated from the pole-zero plot, to within a multiplicative constant
- In the example below, magnitude of numerator equals length of vector drawn from the zero to $z=e^{j\omega}$, magnitude of denominator equals length of vector drawn from pole to $z=e^{j\omega}$
  - Thus, $|H(z)|$ at $z=1$ is $|H(z)|\approx 5(2)/(0.1)\approx 100$
  - Similarly, $|H(z)|$ at $z=j$ is $|H(z)|\approx 5(\sqrt{2})/(\sqrt{2})\approx 5$
  - $H(-1)=0$

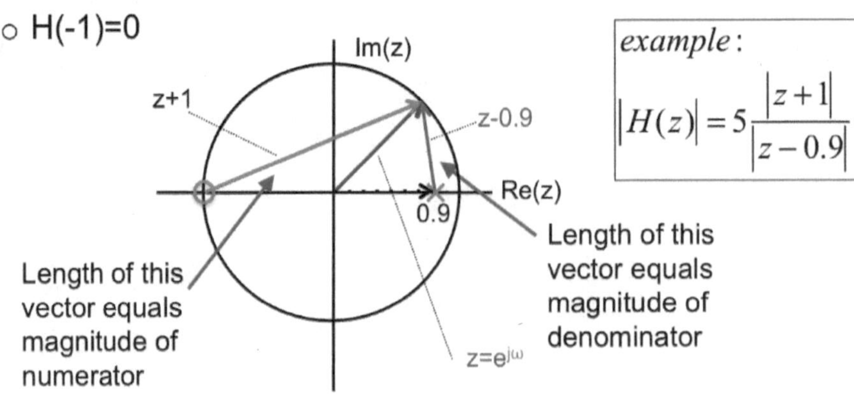

Freq. Response from Poles & Zeroes

- Frequency response of a filter from simulation
- Previous slide graphical estimates were
  - o |H(z)| at z=1 is |H(z)|≈5(2)/(0.1)≈100
  - o |H(z)| at z=j is |H(z)|≈5(√2)/(√2)≈5
  - o H(-1)=0

$$|H(z)| = 5\frac{|z+1|}{|z-0.9|}$$

|H(z)|=5.26 @ ω=π/2

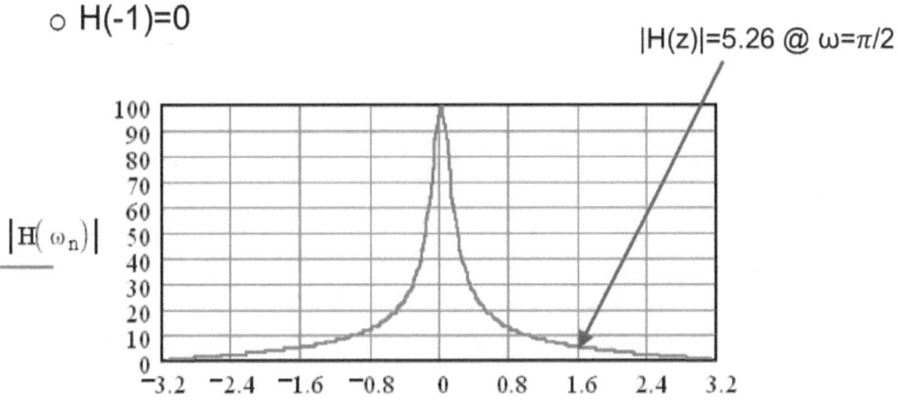

Filter Transformations (A.G. Constantinides)

Transformation	Replace lowpass z^{-1} with:	Variables
Lowpass to Lowpass	$\frac{z^{-1}-a}{1-az^{-1}}$	$a=\frac{\sin[(\omega_{cold}-\omega_{cnew})/2]}{\sin[(\omega_{cold}+\omega_{cnew})/2]}$
Lowpass to Highpass	$-\frac{z^{-1}+a}{1+az^{-1}}$	$a=\frac{\cos[(\omega_{cold}+\omega_{cnew})/2]}{\cos[(\omega_{cold}-\omega_{cnew})/2]}$
Lowpass to Bandpass	$-\left(\frac{z^{-2}-a_1z^{-1}+a_2}{a_2z^{-2}-a_1z^{-1}+1}\right)$	$a_1=2\alpha\beta/(\beta+1)$ $a_2=(\beta-1)/(\beta+1)$ $\alpha=\frac{\cos[(\omega_{uppernew}+\omega_{lowernew})/2]}{\cos[(\omega_{uppernew}-\omega_{lowernew})/2]}$ $\beta=$ $\cot[(\omega_{uppernew}-\omega_{lowernew})/2]\tan\left(\frac{\omega_{cold}}{2}\right)$

9 NOVEL APPLICATIONS

The lecture notes in this chapter present recent novel applications of digital signal processing.

Digital Impedance

Digital Impedance and Non-Foster Circuits

- Can analog RLC circuit be directly replaced by digital?
  - Yes! …but why?
- To implement exotic (non-Foster) components
  - Negative capacitors, negative inductors, negative resistors
  - For many years, analog implementations have had issues with potential instability
- Solution: Digital Non-Foster Circuits
  - Digital is better …400 GHz at 28 nm:
  - Non-Foster yields bandwidth improvement
  - Focus here: negative Capacitors, negative Inductors
  - Digital Discrete-Time non-Foster circuits (ISCAS 2015):
  - *… but still must be stable*

Digital Implementation of Analog RLC

- Often, DSP has replaced analog input/output systems
- Typically, start with analog/continuous-time system

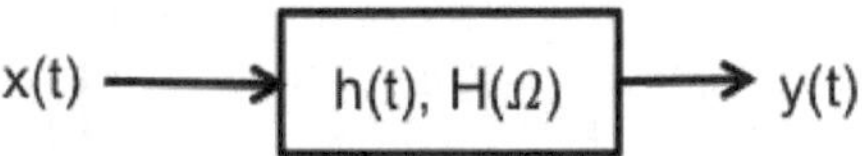

- Replace analog system with digital system

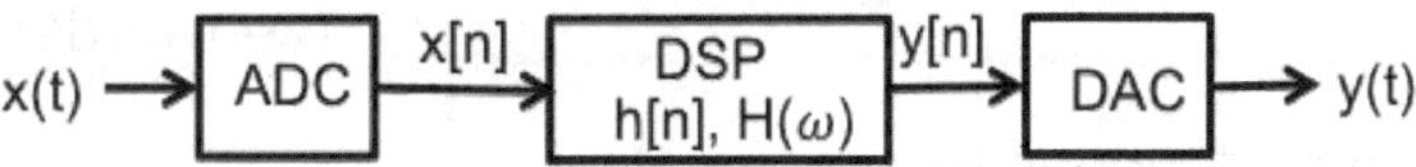

- **A new approach: Digital Implementation of RLC**

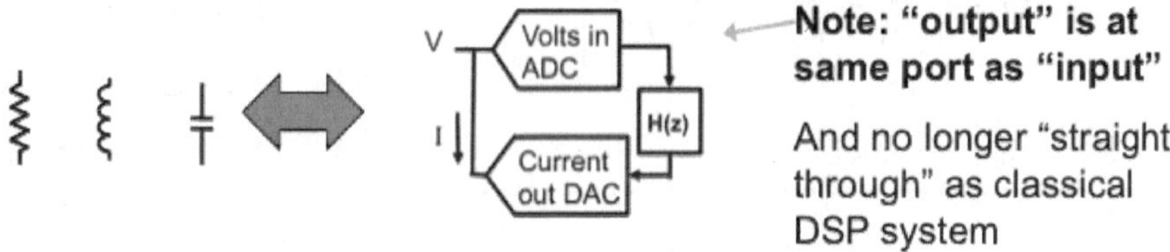

What is Digital RLC?

Simple Example: Digital Resistor

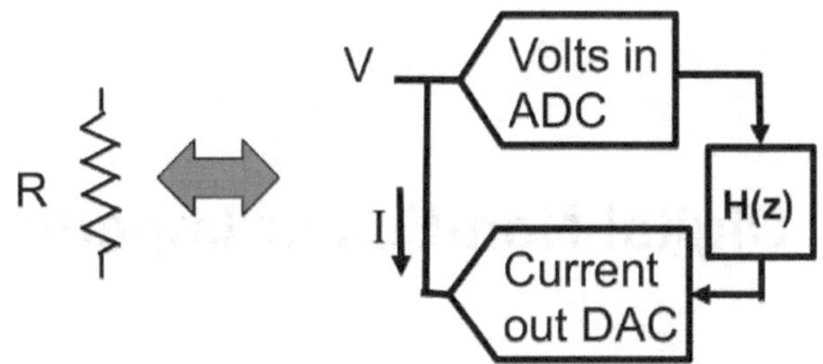

•Measure voltage V
•Set current I
•Let H(z) = 1/R
•So, DAC current: I = V/R

… yields world's most expensive resistor!
… but is tunable

General Theory

- Measure v(t) ADC: v[n] = v(nT)
- Calculate i[n] H(z): i[n] = v[n]*h[n]
- Set i(t) DAC: i(nT) = i[n]

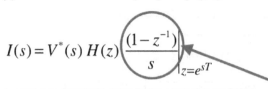

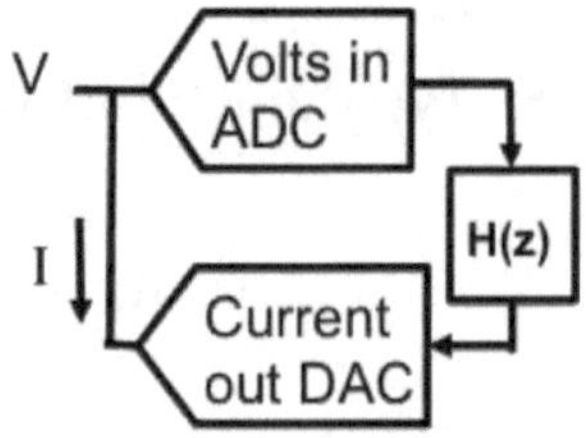

$$I(s) = V^*(s)\,H(z)\left.\frac{(1-z^{-1})}{s}\right|_{z=e^{sT}}$$

(zero-order hold)

Input impedance is then:

$$Z(s) = \frac{V(s)}{I(s)} \approx \left.\frac{sT}{\left(1-z^{-1}\right)H(z)}\right|_{z=e^{sT}}$$

$$V^*(s) = \sum v(nT)e^{-nsT}$$
$$= \sum V(s-n\omega_0)/T$$

(Starred Transform)

Digital Non-Foster Impedance

Digital Negative Capacitor Example

V
Volts in ADC
H(z)
I
Current out DAC

Capacitor:

$$i(t) = C\frac{dv(t)}{dt} \approx C\frac{v[n]-v[n-1]}{T}$$

Taking z-transform:

$$I(z) = C\left(1-z^{-1}\right)V(z)/T = H(z)V(z)$$

Yields the transfer function H(z):

$$H_C(z) = C\left(1-z^{-1}\right)/T$$

The impedance is then:

$$Z_C(s) \approx \left.\frac{sT}{\left[\left(1-z^{-1}\right)H(z)\right]}\right|_{z=e^{sT}} = \left.\frac{sT^2}{C\left(1-z^{-1}\right)^2}\right|_{z=e^{sT}} \approx \frac{1}{sC}$$

Why Non-Foster Circuits?

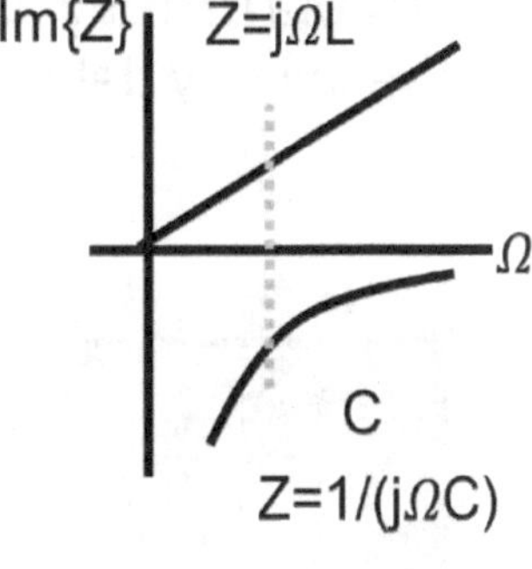

- Of particular interest:
  - Negative capacitors
  - Negative inductors
- Useful for many applications
  - Wideband small antenas
  - Wideband metamaterials
  - Superluminal lines
  - Cancel Parasitics

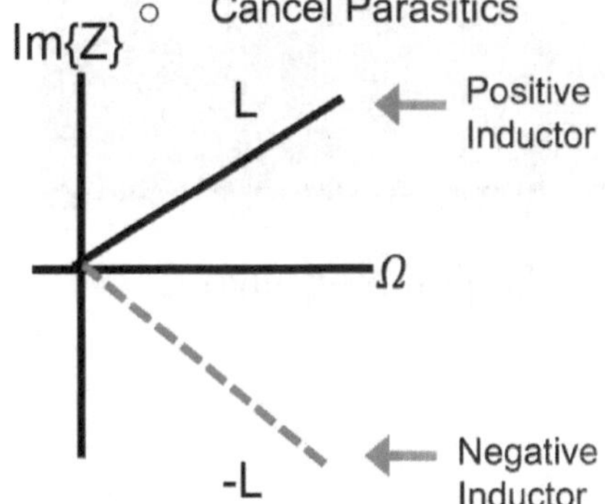

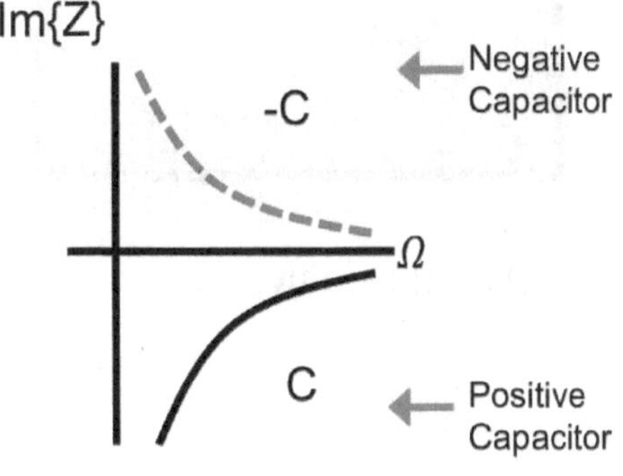

Digital Negative Capacitor Simulation Results

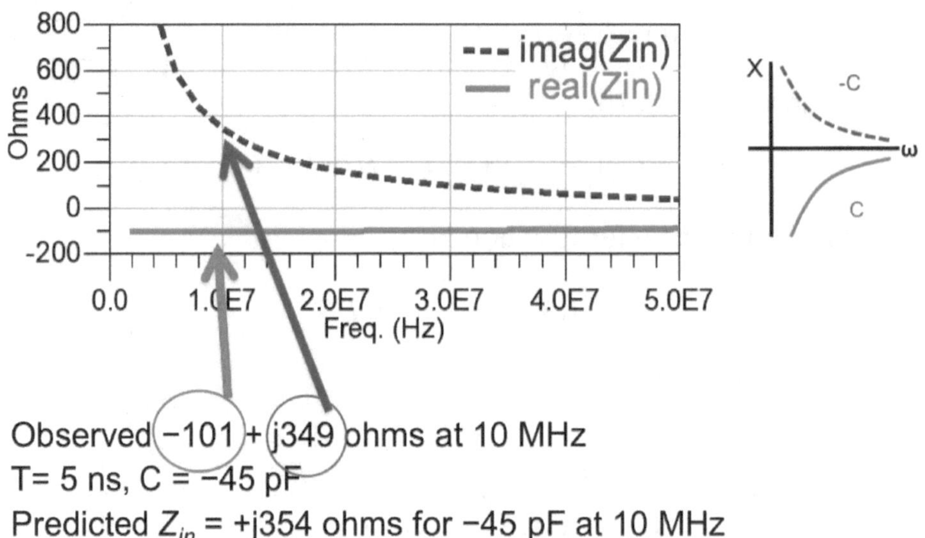

Preliminary Data: Positive and Negative Cap

- Preliminary data (alternative topology)

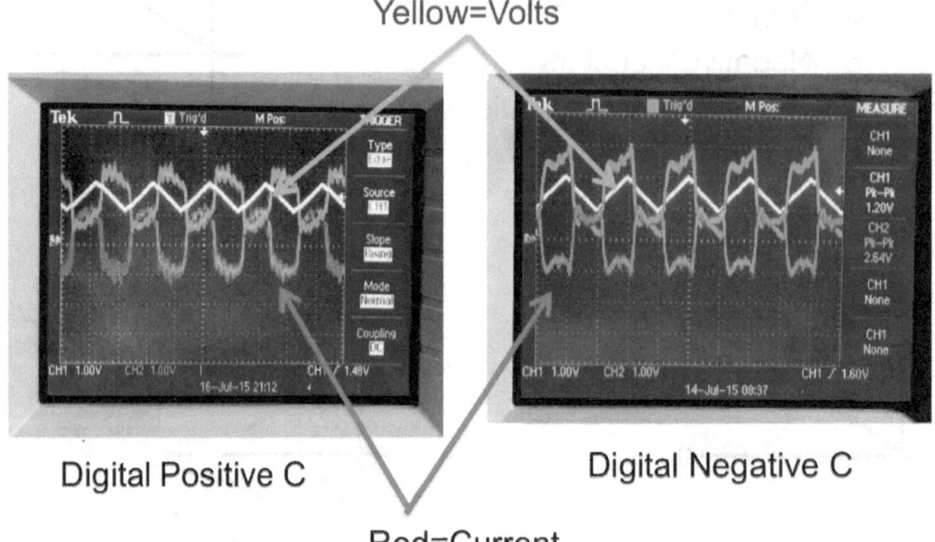

Digital Negative Inductor

- To implement a digital inductor, proceed in similar fashion
- For any inductor,
$$i(t) = i(0) + \int v(t)dt / L.$$
  - The current may be approximated by using a discrete-time accumulator approximation
$$i(0) + \int v(t)dt / L \approx i[0] + \sum v[n]T / L,$$
and setting $$i[n] = i[n-1] + Tv[n] / L.$$
  - This is the discrete-time current, where *L* is the desired inductance and *T* is the sampling period.
  - Taking the z-transform,
$$I(z) = TV(z) / \left(L - z^{-1}L\right) = H(z)V(z).$$

Digital Negative Inductor

- The transfer function becomes
$$H_L(z) = \frac{T}{L\left(1 - z^{-1}\right)}.$$
- The impedance becomes
$$Z_L(s) = \left. \frac{sT}{\left[\left(1 - z^{-1}\right)H_L(z)\right]} \right|_{z=e^{sT}} \approx sL$$

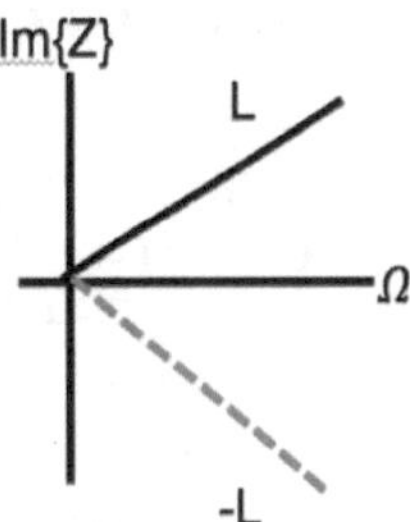

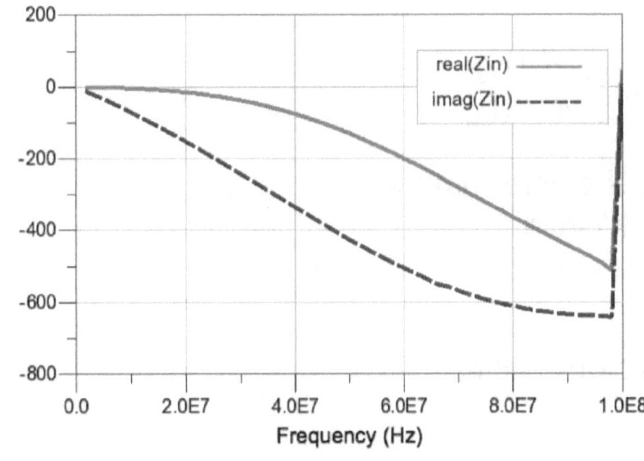

- Design target
- L = −1 μH at T = 5 ns, predicted Z = −j62.8 ohms at 10 MHz
- Simulated

Z = −3.5 + -j70 at 10 MHz

Thevenin Approach

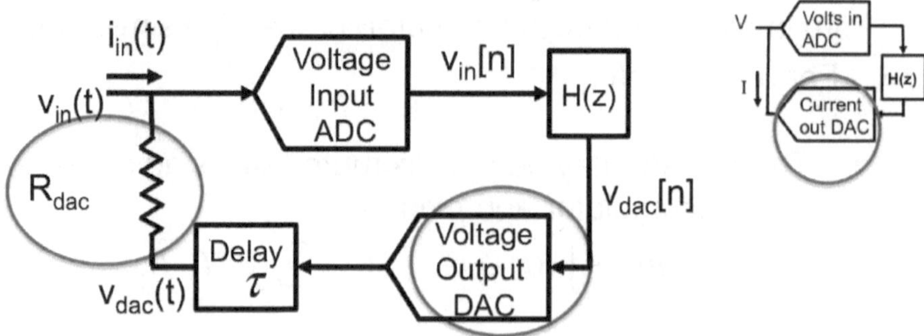

- Voltage DAC plus R_{dac} is a Thevenin source
- Could do simple *Thevenin-to-Norton* transformation
- Better approach: incorporate R_{dac} into H(z) design
- Time-delay latencies are also modeled
- Latency caused by ADC conversion and computation time

RC Stability Analysis Theory

- The corresponding discrete-time transfer function is

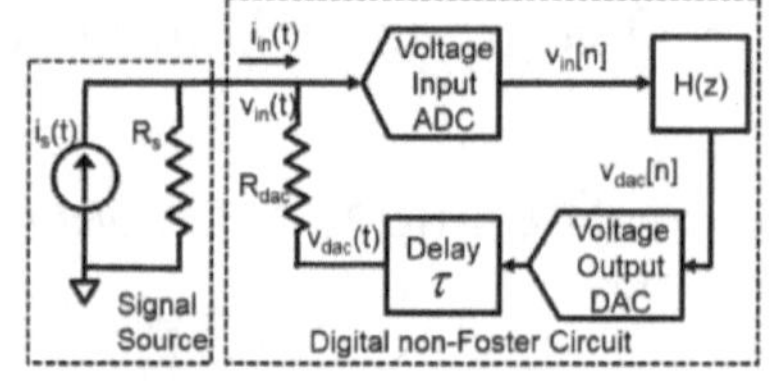

$$G_{RC}(z) = \frac{I_{in}(z)}{I_s(z)} = \frac{-R_e z^{-\lambda} H_{RC}(z) / R_{dac}}{1 - R_e z^{-\lambda} H_{RC}(z) / R_{dac}}$$

$$with: R_e = R_s \,||\, R_{dac} = \frac{R_s R_{dac}}{R_s + R_{dac}}$$

Where the system is stable with the poles inside the unit circle,

λ is the latency in clock cycles

RC Stability Analysis with Varying Latency

Pole and zero locations with C = -8nF and Rser = 50Ω while latency is varied from 1 to 5 clock cycles

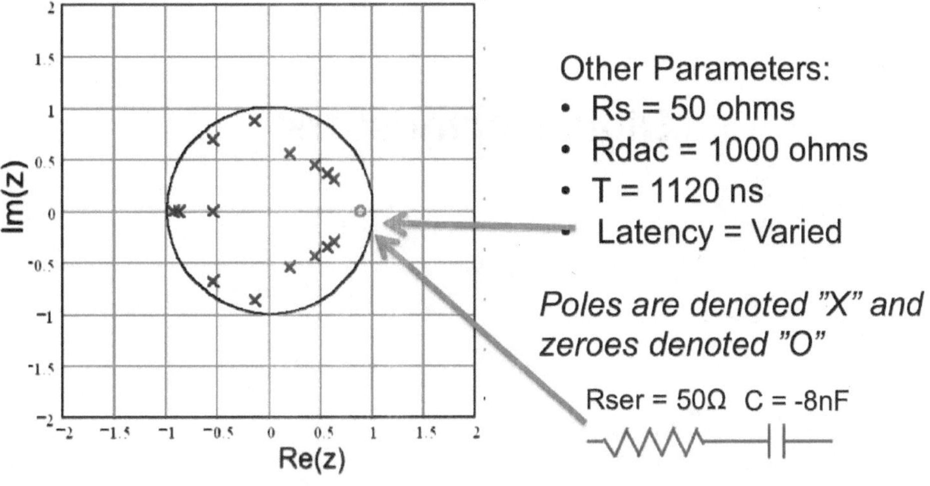

Stable up to Latency = 4

Thevenin Approach: General Theory

•Measure v(t)	ADC:	$v[n] = v(nT)$
•Calculate $v_{dac}[n]$	H(z):	$v_{dac}[n] = v[n]*h[n]$
•Set $v_{dac}(t)$	DAC:	$v_{dac}(nT) = v_{dac}[n]$

$i_{in}(t)$, $v_{in}(t)$, R_{dac}, $v_{dac}(t)$, Voltage Input ADC, $v_{in}[n]$, H(z), $v_{dac}[n]$, Voltage Output DAC, Delay τ

$$V_{dac}(s) = \left.\frac{V^*(s)H(z)(1-z^{-1})e^{-s\tau}}{s}\right|_{z=e^{sT}}$$

$$I(s) \approx \frac{V_{in}(s) - V_{dac}(s)}{R_{dac}}$$

$$V_{dac}(z) = H(z)V_{in}(z)$$

Input impedance is then:

$$Z(s) = \frac{V(s)}{I(s)} \approx \left.\frac{sTR_{dac}}{\left[sT - H(z)\left(1-z^{-1}\right)e^{-s\tau}\right]}\right|_{z=e^{sT}}$$

$$V^*(s) = \Sigma v(nT)e^{-nsT}$$

$$= \Sigma V(s - n\omega_0)/T$$

(Starred Transform)

Digital Memristors

Memristors

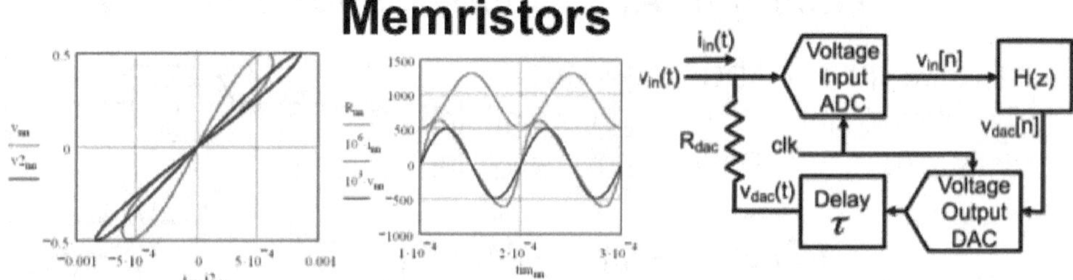

- Memristor is a non-linear impedance
- Implemented using same configuration as before
- Memristor resistance depends on flux or charge
- Flux φ is integral of voltage, charge q is integral of current

$$i[n] = v[n]G(\varphi) = K_G \frac{v[n]}{\varphi[n]} \quad and \quad \varphi[n] = \varphi_0 + T_C \sum_{\alpha=1}^{n} v[\alpha]$$

$$where: \quad R[n] = \frac{1}{G(\varphi)} \text{ is the resistance}$$

- Memristor has memory
- Applications: memory cells, neural computing

10 CLASSIFIERS

This lecture notes in this chapter present a brief overview of basic concepts in the design of a variety of different classifiers.

Classifiers

- Often, the last step in a signal processing system is to make some sort of decision
- Classifiers are used to implement such decision functions
- There are many different types of classifiers:
  - Gaussian optimum threshold
  - Multivariate Gaussian
  - Bayesian Classifier
  - Linear Discriminant Classifier
  - K-Means Classifier
  - K-Nearest Neighbor Classifier
  - Minimum-Distance Classifier
  - Supervised / Unsupervised

Single-Feature (Scalar) Threshold Classifier

- Suppose a signal processing system with 1 feature
  - Scalar feature g

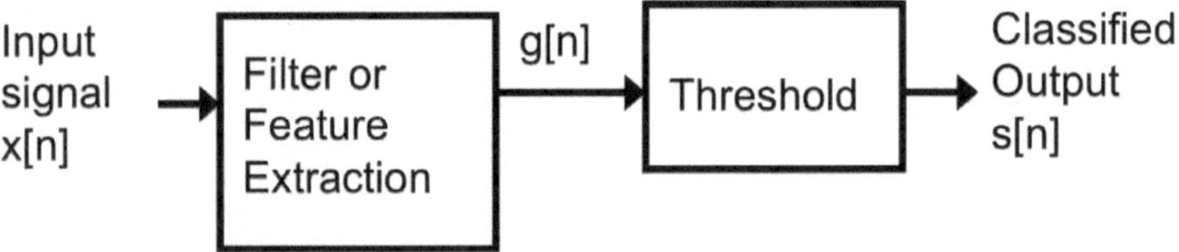

Optimum Single-Feature Threshold

- Optimum Bayes classifier: use maximum likelihood ratio
    - $P(g|C_1)$ = probability density of g for class C_1
    - P_{C1} = *a priori* probability of class C_1
    - g = observed scalar feature value
  - The optimum Bayes classifier is
    - Choose class $s[n]=C_a$ when
      $P(g|C_a)\ P_{ca} \geq P(g|C_b)\ P_{Cb}$ for all C_b (choose most likely)

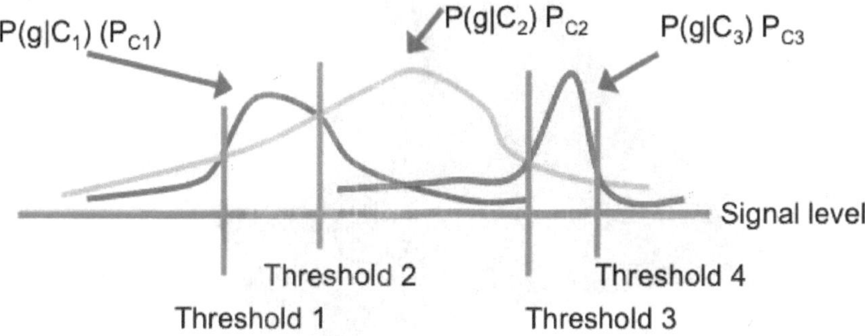

Vector-Classifier

Suppose a signal processing system with k features

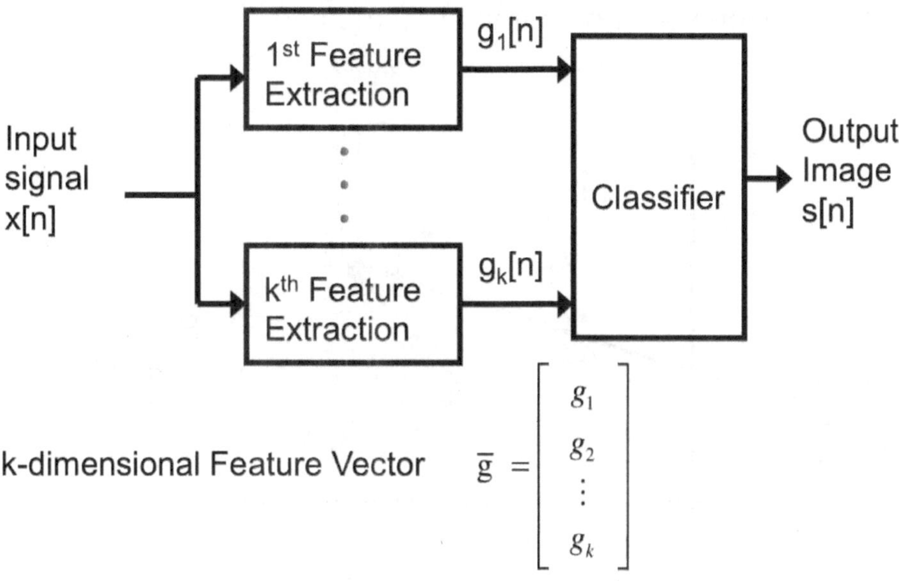

$$\bar{g} = \begin{bmatrix} g_1 \\ g_2 \\ \vdots \\ g_k \end{bmatrix}$$

Multivariate Gaussian Example

- Bayes classifier: choose most likely class
- Example below:
  - k=2 dimensional vector, with 2 classes shown

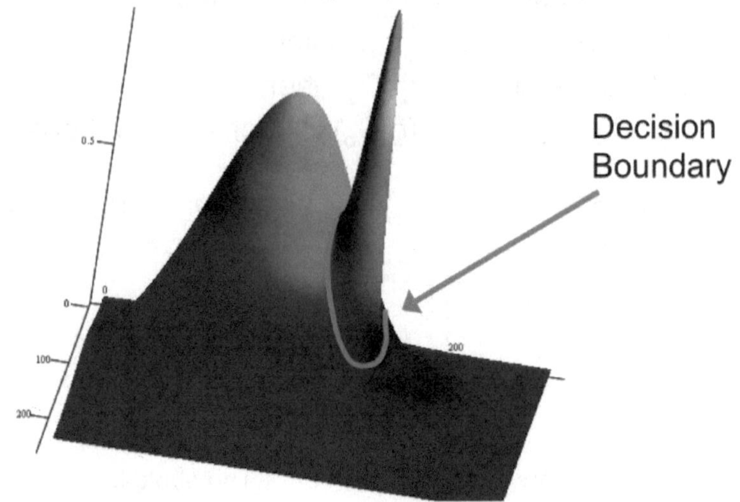

Linear Discriminant Classifier

- Simple clustering: Linear boundary (line or hyperplane)
- Distribution-free (non-parametric) method

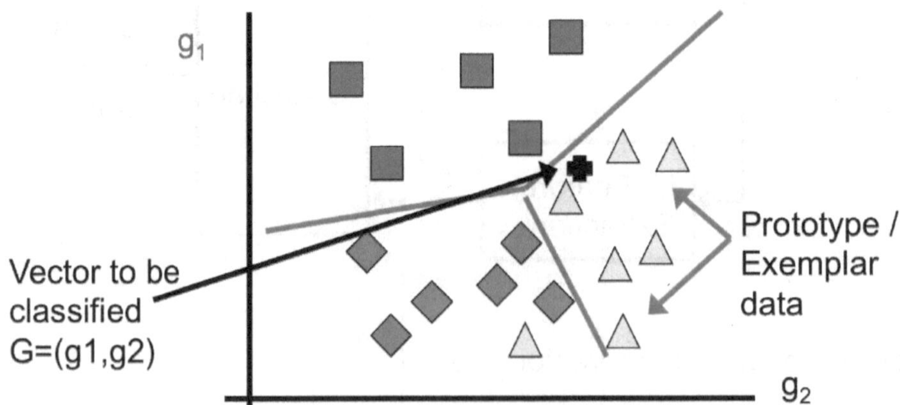

K-Means Classifier

- K-means clustering (minimum mean distance classifier):
  - o Compute cluster centers,
  - o Classifier assigns to class with closest center (mean)

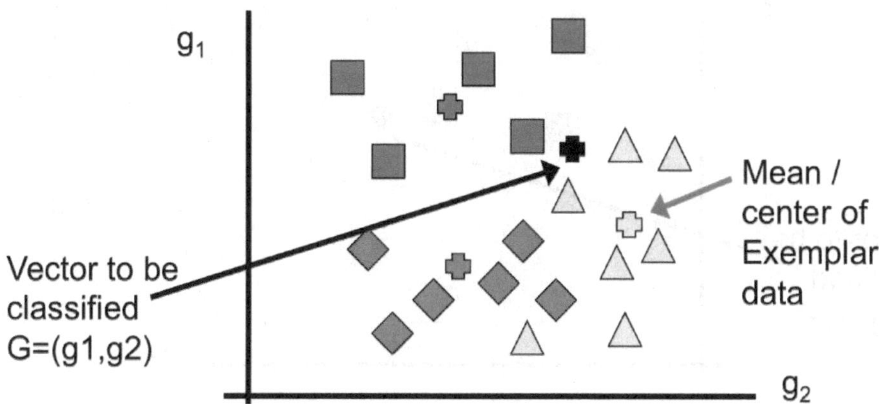

K-Nearest Neighbor Classifier (K=2)

- K-nearest neighbor:
  - o Find the k nearest neighbors of the vector G,
  - o Assigns to majority class of the k neighbors

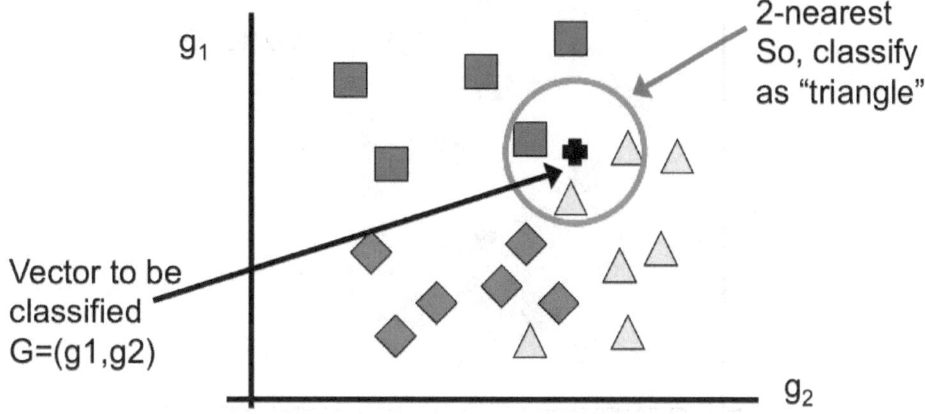

K-Nearest Neighbor Classifier (K=3)

- K-nearest neighbor:
  - Find the k nearest neighbors of the vector G,
  - Assigns to majority class of the k neighbors

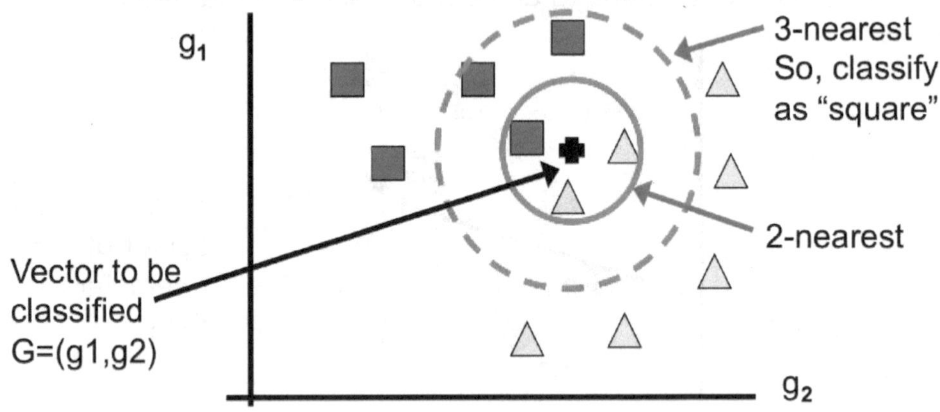

Minimum-Distance Classifier

- Minimum-distance Classifier: (1-nearest neighbor)
  - Find the nearest neighbor of the vector G,
  - Assign to that class

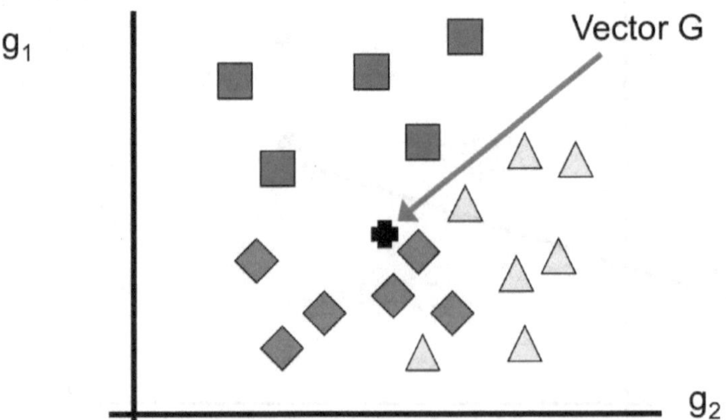

Unsupervised Classifier

- As a first step, an unsupervised classifier would look at the data and organize it into a number of classes based on "clustering" of points in feature space
- Then, any of the prior methods, such as nearest neighbor, could be used to assign a vector to a class

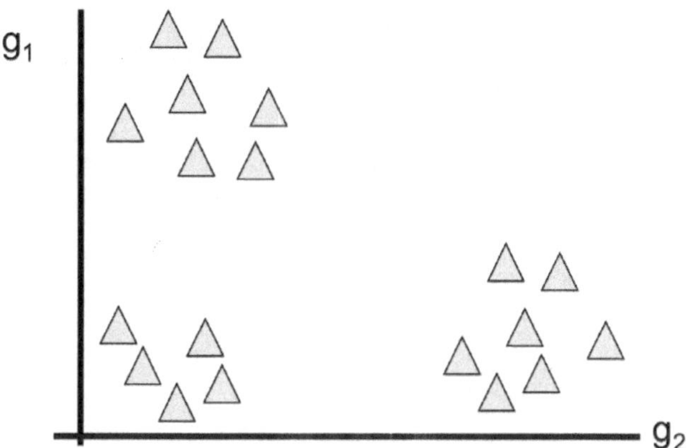

Classifier for Binary Data

- A common problem in communication systems is to determine whether noisy binary data is a "1" or "0"
- A simple threshold may be used in some cases
- However, more complex classifiers may reduce error rates
- For example, noisy data illustrated below:

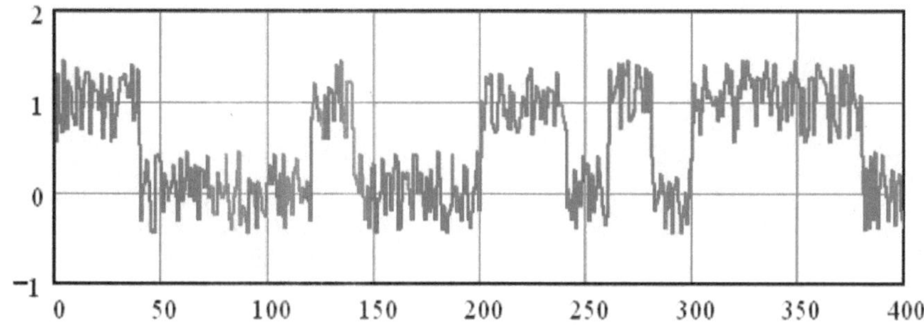

11 APPENDIX

Z-transform Pairs

Discrete-time Function	z-transform
$\delta[n]$	1
$u[n]$	$\frac{z}{z-1};\ \lvert z\rvert > 1$
$nu[n]$	$\frac{z}{(z-1)^2};\ \lvert z\rvert > 1$
$a^n u[n]$	$\frac{z}{z-a};\ \lvert z\rvert > \lvert a\rvert$
$-a^n u[-n-1]$	$\frac{z}{z-a};\ \lvert z\rvert < \lvert a\rvert$
$na^n u[n]$	$\frac{az}{(z-a)^2};\ \lvert z\rvert > \lvert a\rvert$
$\cos(\omega_0 n)u[n]$	$\frac{z^2 - z\cos(\omega_0)}{z^2 - 2z\cos(\omega_0)+1};\ \lvert z\rvert > 1$
$\sin(\omega_0 n)u[n]$	$\frac{z\sin(\omega_0)}{z^2 - 2z\cos(\omega_0)+1};\ \lvert z\rvert > 1$

Trigonometry Identities

$$cos(A)\ cos(B) = 0.5\ cos(A - B) + 0.5\ cos(A + B)$$
$$sin(A)\ cos(B) = 0.5\ sin(A - B) + 0.5\ sin(A + B)$$
$$sin(A) sin(B) = 0.5\ cos(A - B) - 0.5\ cos(A + B)$$
$$\cos^2(A) = 0.5 + 0.5\cos(2A)$$
$$\sin^2(A) = 0.5 - 0.5\cos(2A)$$
$$a\cos(x) + b\sin(x) = \sqrt{a^2+b^2}\cos(x + \arctan(-b/a))$$
$$e^{j\theta} = \cos(\theta) + j\ \sin(\theta)$$
$$e^{j\theta} + e^{-j\theta} = 2\cos(\theta)$$
$$e^{j\theta} - e^{-j\theta} = 2j\sin(\theta)$$

Fourier Transform in "f"

Fourier Transform Pairs $X(f) = \int_{-\infty}^{\infty} x(t)e^{-j2\pi ft}dt \qquad x(t) = \int_{-\infty}^{\infty} X(f)e^{j2\pi ft}df$

$\delta(t) \leftrightarrow 1$		$\cos(2\pi f_0 t) \leftrightarrow 0.5(\delta(f+f_0)+\delta(f-f_0))$
$1 \leftrightarrow \delta(f)$		$\sin(2\pi f_0 t) \leftrightarrow 0.5j(\delta(f+f_0)-\delta(f-f_0))$
$u(t) \leftrightarrow \frac{1}{2}\delta(f) + \frac{1}{j2\pi f}$		$sgn(t) \leftrightarrow \frac{1}{j\pi f}$
$\Pi(t/\tau) \leftrightarrow \tau \cdot sinc(\pi f \tau)$		$2B\, sinc(2\pi Bt) \leftrightarrow \Pi(f/(2B))$
$\Delta\left(\frac{t}{\tau}\right) \leftrightarrow \frac{\tau}{2} \cdot sinc^2(\pi f\tau/2)$		$B\, sinc^2(\pi Bt) \leftrightarrow \Delta(f/(2B))$
$e^{j2\pi f_0 t} \leftrightarrow \delta(f-f_0)$		$e^{-at}u(t) \leftrightarrow \frac{1}{a+j2\pi f}$
$e^{-t^2/(2\sigma^2)} \leftrightarrow \sigma\sqrt{2\pi}\, e^{-2(\sigma\pi f)^2}$		$e^{-a\lvert t\rvert} \leftrightarrow \frac{2a}{a^2+(2\pi f)^2}$

Fourier Transform Properties

$g(t)e^{j2\pi f_0 t} \leftrightarrow G(f-f_0)$		$g(t-t_0) \leftrightarrow G(f)e^{-j2\pi t_0 f}$
$g(at) \leftrightarrow \frac{1}{\lvert a\rvert}G\left(\frac{f}{a}\right)$		$G(t) \leftrightarrow g(-f)$
$g(t) * h(t) \leftrightarrow G(f)H(f)$		$g(t)h(t) \leftrightarrow G(f) * H(f)$
$\frac{dg(t)}{dt} \leftrightarrow j2\pi f\, G(f)$		$\int_{-\infty}^{t} g(\alpha)d\alpha \leftrightarrow \frac{G(f)}{j2\pi f} + G(0)\delta(f)/2$
		$\int_{-\infty}^{\infty} \lvert g(t)\rvert^2 dt = \int_{-\infty}^{\infty} \lvert G(f)\rvert^2 df$

Fourier Transform in "Ω"

Fourier Transform Pairs $X(\Omega) = \int_{-\infty}^{\infty} x(t)e^{-j\Omega t}dt$ $\quad x(t) = \frac{1}{2\pi}\int_{-\infty}^{\infty} X(\Omega)e^{j\Omega t}d\Omega$

$\delta(t) \leftrightarrow 1$		$\cos(\Omega_0 t) \leftrightarrow \pi(\delta(\Omega+\Omega_0)+\delta(\Omega-\Omega_0))$
$1 \leftrightarrow 2\pi\delta(\Omega)$		$\sin(\Omega_0 t) \leftrightarrow j\pi(\delta(\Omega+\Omega_0)-\delta(\Omega-\Omega_0))$
$u(t) \leftrightarrow \pi\delta(\Omega) + \frac{1}{j\Omega}$		$sgn(t) \leftrightarrow \frac{2}{j\Omega}$
$\Pi(t/\tau) \leftrightarrow \tau \cdot sinc(\Omega\tau/2)$		$W\, sinc(Wt) \leftrightarrow \pi\, \Pi(\Omega/(2W))$
$\Delta\left(\frac{t}{\tau}\right) \leftrightarrow \frac{\tau}{2} \cdot sinc^2(\Omega\tau/4)$		$W\, sinc^2(Wt/2) \leftrightarrow 2\pi\Delta(\Omega/(2W))$
$e^{j\Omega_0 t} \leftrightarrow 2\pi\delta(\Omega-\Omega_0)$		$e^{-at}u(t) \leftrightarrow \frac{1}{a+j\Omega}$
$e^{-t^2/(2\sigma^2)} \leftrightarrow \sigma\sqrt{2\pi}\, e^{-\sigma^2\Omega^2/2}$		$e^{-a\lvert t\rvert} \leftrightarrow \frac{2a}{a^2+\Omega^2}$

Fourier Transform Properties

$g(t)e^{j\Omega_0 t} \leftrightarrow G(\Omega-\Omega_0)$		$g(t-t_0) \leftrightarrow G(\Omega)e^{-jt_0\Omega}$
$g(at) \leftrightarrow \frac{1}{\lvert a\rvert}G\left(\frac{\Omega}{a}\right)$		$G(t) \leftrightarrow 2\pi g(-\Omega)$
$g(t) * h(t) \leftrightarrow G(\Omega)H(\Omega)$		$g(t)h(t) \leftrightarrow \frac{1}{2\pi}G(\Omega) * H(\Omega)$
$\frac{dg(t)}{dt} \leftrightarrow j\Omega\, G(\Omega)$		$\int_{-\infty}^{t} g(\alpha)d\alpha \leftrightarrow \frac{G(\Omega)}{j\Omega} + \pi G(0)\delta(\Omega)$
		$\int_{-\infty}^{\infty} \lvert g(t)\rvert^2 dt = \frac{1}{2\pi}\int_{-\infty}^{\infty} \lvert G(\Omega)\rvert^2 d\Omega$

Table of Q(α)

α	Q(α)
0	0.5000
0.5	0.3085
1.0	0.1587
1.5	0.0668
2.0	0.0228
2.5	0.0062
3.0	0.0014

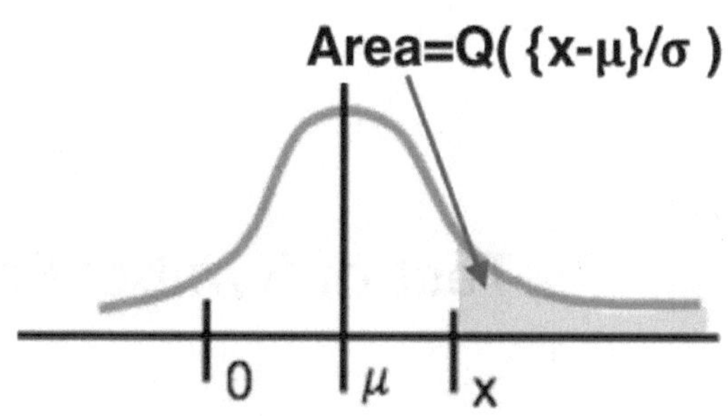

Matrix Identities

$$\left(\bar{\bar{A}}\,\bar{\bar{B}}\right)^{-1} = \bar{\bar{B}}^{-1}\,\bar{\bar{A}}^{-1}$$

$$\left(\bar{\bar{A}}\,\bar{\bar{\mathrm{B}}}\right)^{H} = \bar{\bar{B}}^{H}\,\bar{\bar{A}}^{H}$$

$$\left(\bar{\bar{A}}+\bar{\bar{B}}\right)^{H} = \bar{\bar{A}}^{H}+\bar{\bar{B}}^{H}$$

$$\left(\bar{\bar{A}}\,\bar{\bar{B}}\right)^{T} = \bar{\bar{B}}^{T}\,\bar{\bar{A}}^{T}$$

$$\left(\bar{\bar{A}}\,\bar{\bar{B}}\right)^{T} = \bar{\bar{B}}^{T}\,\bar{\bar{A}}^{T}$$

$$\left(\bar{\bar{A}}^{H}\right)^{-1} = \left(A^{-1}\right)^{H}$$

$$\left(A^{T}\right)^{-1} = \left(A^{-1}\right)^{T}$$

Quadratic formula:

$$ax^2+bx+c=0 \quad \Rightarrow \quad x=\frac{-b\pm\sqrt{b^2-4ac}}{2a}$$

Test of Symbol Fonts

- Arial symbols:
- ©®Ωℑ𝔎𝓕𝒁⊕××⊗∩∪⊕⊗≈∫⋀≥≦≤∆∇ ∀ ∞
- →→⟹⇔×÷±∓≈≠∘ ▪•●∗·√∑ ∫∠ | ∮ ∯ ∮
- *αβγδεζηθικλμνξοπρςστυφχψω ∂ϵϑϰϕϱϖ***A**
- *ΑΒΓΔΕΖΗΘΙΚΛΜΝΞΟΠΡΘΣΤΥΦΧΨΩ* ∇

END

Made in United States
Troutdale, OR
01/03/2025